Springer Theses

Recognizing Outstanding Ph.D. Research

For further volumes:
http://www.springer.com/series/8790

Aims and Scope

The series "Springer Theses" brings together a selection of the very best Ph.D. theses from around the world and across the physical sciences. Nominated and endorsed by two recognized specialists, each published volume has been selected for its scientific excellence and the high impact of its contents for the pertinent field of research. For greater accessibility to non-specialists, the published versions include an extended introduction, as well as a foreword by the student's supervisor explaining the special relevance of the work for the field. As a whole, the series will provide a valuable resource both for newcomers to the research fields described, and for other scientists seeking detailed background information on special questions. Finally, it provides an accredited documentation of the valuable contributions made by today's younger generation of scientists.

Theses are accepted into the series by invited nomination only and must fulfill all of the following criteria

- They must be written in good English.
- The topic of should fall within the confines of Chemistry, Physics and related interdisciplinary fields such as Materials, Nanoscience, Chemical Engineering, Complex Systems and Biophysics.
- The work reported in the thesis must represent a significant scientific advance.
- If the thesis includes previously published material, permission to reproduce this must be gained from the respective copyright holder.
- They must have been examined and passed during the 12 months prior to nomination.
- Each thesis should include a foreword by the supervisor outlining the significance of its content.
- The theses should have a clearly defined structure including an introduction accessible to scientists not expert in that particular field.

Ricardo A. S. Ramalho

Building the Cape Verde Islands

Doctoral Thesis accepted by the University of Bristol, Bristol, United Kingdom

Springer

Author
Dr. Ricardo A. S. Ramalho
School of Earth Sciences
University of Bristol
Queen's Road, Bristol
BS8 1RJ
UK
e-mail: Ric.Ramalho@bristol.ac.uk

Supervisors
Profs. George Helffrich and Derek Vance
School of Earth Sciences
University of Bristol
Queen's Road, Bristol
BS8 1RJ
UK

ISSN 2190-5053 e-ISSN 2190-5061

ISBN 978-3-642-19102-2 e-ISBN 978-3-642-19103-9

DOI 10.1007/978-3-642-19103-9

Springer Heidelberg Dordrecht London New York

Cover design: eStudio Calamar, Berlin/Figueres

Printed on acid-free paper

Springer is part of Springer Science+Business Media (www.springer.com)

I would like to dedicate this work to Professor António Serralheiro and Dr. Luís Celestino Silva, the true pioneers of Cape Verdean geology; I would also would like to dedicate this work to Professor António Ribeiro, who was always a source of inspiration and who transmits his dedication to science as a "contagious disease".

Parts of this thesis have been published in the following journal articles:

Ramalho, R., Helffrich, G., Schmidt, D. N., & Vance, D. (2010). Tracers of uplift and subsidence in the Cape Verde Archipelago. *Journal of the Geological Society, 167*, 519–538.

Ramalho, R., Helffrich, G., Cosca, M., Vance, D., Hoffmann, D., & Schmidt, D. N. (2010). Vertical movements of ocean island volcanoes: insights from a stationary plate. *Marine Geology, 275*, 84–95.

Supervisor's Foreword

Hotspots are enigmatic surface features that are not easily explained in the framework of plate tectonics. Investigating their origin is the goal of this thesis, using field evidence collected in the Cape Verde Islands, a prominent hotspot archipelago in the eastern Atlantic Ocean. The approach taken is to document uplift of the islands relative to sea level and use the uplift features to test various models of hotspot development. Island uplift is thought to arise from the growth of the anomalously shallow sea floor on which the islands rest, known as the bathymetric swell, which is characteristic of hotspots.

The work comprises a geological summary and detailed mapping of palaeo sea-level markers on Cape Verde. Isotopic dating of the markers shows that uplift on the islands over the last 6 Ma is up to 400 m, and that the uplift chronology varies among islands. Two processes act to raise the Cape Verde Islands. The dominant process is one that is local to individual islands. The regional, swell-related component is smaller, and possibly episodic. The observations are strong constraints on swell development and on hotspot models.

This work provides an extensive field data set for the testing of future models of hotspot swell development. The study also shows how the traditional techniques of field geology can make a key contribution to contemporary geophysical research. I hope that the methods used here might stimulate investigations of other hotspot-influenced islands and generate a wider set of documented behaviours against which hotspot ideas might be tested.

February 2011 Prof. George Helffrich and Derek Vance

Preface

Research is not an end in itself: to communicate results, methodologies and theories is so fundamental that publishing became an integral part of the research process. The global human progress is intrinsically linked not only to the *advancement* of knowledge, but also to the *diffusion* of knowledge. The dissemination of recently-acquired knowledge among stakeholders is also crucial to attract interest and, lets face it, funding. Likewise, academic researchers have the responsibility towards their sponsors and funding bodies to show the results of their research activities and their contribution to the advancement of the particular topics they are working on. The publication of new scientifc findings is based on a peer-review process so others may comment on our work and positively contribute to the quality of the publications. As it happens, and for the better or for the worse, publication is often considered a metric of performance. Thus, it is useful nowadays, when starting a research project, to set up a publication strategy. Furthermore, the writing and submission of scientific papers is a valuable skill and an integral part of academic training. With this in mind, and following direct advice from my Ph.D. Advisory Committee, the contents of this thesis were as much as possible organised as a set of ready-to-publish materials, some of which are already published. This strategy is well-intentioned and has clear advantages, but some repetition and loss of continuity in this work may arise due to the intrinsic characteristics of a scientific paper (with its own introduction, scientific background, etc.). Not withstanding these caveats, I have tried my best to follow a clear streamline and to ensure the logical connection between subjects. It is my hope that this was achieved, and that this thesis efficiently communicates the core of our findings.

Acknowledgments

All scientific communications are the result of an author's experience and are not possible without the support and contribution of many other people and institutions. This thesis is not an exception and I owe a great deal to many. I must start by expressing my sincere thanks to my supervisors George Helffrich and Derek Vance. This work wouldn't have been possible without their orientation, support and motivation. They have always showed incredible patience to explain me things from the most simple to the most complex and always with an incredible sense of humour. It is hard to express in words my gratitude to both, for making this work a true pleasure. I would also like to express a special word of gratitude to my colleagues in Portugal: first, to António Serralheiro and Luís Celestino Silva, who are always willing to share their solid, life-long knowledge on the volcanostratigraphy of the Cape Verde Islands; secondly, to José Madeira, António Silveira, João Mata, Paulo Fonseca, António Ribeiro and Sérgio Ávila, who always supported me in the best of their abilities. My thanks to Dani Schmidt and Chris Hawkesworth (my Ph.D. Advisory Committee) and to Tim Elliott and Kerry Gallagher (my examiners), all of whom contributed a great deal to the improvement of this work. A great thanks to: Mike Kendall and all the geophysics group—my offices' mates—who countless times helped me and who make the daily work a all lot of fun; to all in the Bristol Isotope Group for their crucial support and guidance through the mysteries of mass spectrometry. Many thanks to Mike Cosca for his willingness to help us with the Ar–Ar dating—his state-of-the-art geochronology allowed the achievement of quality results otherwise difficult to obtain. My appreciation to Dirk Hoffmann and Dave Richards for all their support regarding U-Th dating of corals. Also a special word to Bruno Faria, Ed Harms and the Jardim family, who gave a great support during our field trips to Cape Verde, and afterwards. My acknowledgements to Patrick Orr for the making of thin sections. My thanks to John Todd, Gary Nichols, Nick Minter and Haydon Bailey for their support with palaeontological and sedimentological features. I thoroughly thank Fundação para a Ciência e Tecnologia (FCT) for the funding of the Ph.D. scholarship SFRH/BD/24835/2005, co-financed by POPH/FSE. I truly recognise the role of this institution in my personal development. My recognition to the School of Earth Sciences, who supported my development through the Transferable Skills Program and other initiatives, and to Laboratório de Tectonofísica e Tectónica Experimental (LATTEX) for the making of some of the thin sections used in this work. Last but not least, a very special thanks to my partner Jessica and to my family who always supported me, unconditionally. Thank you with all my heart.

Contents

Chapter 1
Introduction

1.1 Introduction

The principle of isostasy describes the state to which the Earth's crust and mantle tend in the absence of disturbing forces [1]. However, this is an idealised state. The mass transfer to and over the surface of the planet—through mountain building, volcanism, erosion and sedimentation, ice sheet loading, etc—disturbs isostasy and generates vertical movements that affect the Earth's surface often without reaching equilibrium [1]. An understanding of the causes and effects of such movements gives insight into the processes of interaction between the planet's interior and surface. In a similar fashion, the study of isostasy plays a key role in the characterisation of the physical properties of our planet's lithosphere.

Plate tectonics and hotspot activity are probably the main agents affecting isostatic equilibrium to which our planet tends. Plate tectonics creates the Earth's primary relief through plate collision, seafloor spreading, arc magmatism, etc., and alters the lithosphere chemically and physically. These processes disrupt the isostatic balance and elicit a feedback from the system. A vertical rebound is thus initiated in order to re-establish equilibrium. Despite the many unanswered questions regarding the driving mechanism, the theory of plate tectonics is now firmly established [2] and our knowledge about tectonically-driven isostasy is fairly robust. Conversely, our knowledge regarding hotspots and their driving mechanisms is still in its infancy. The simple fact that the plume hypothesis has failed to provide a single unifying model that explains and predicts the diversity of features exhibited by hotspots attests to our poor knowledge regarding hotspot governing dynamics. Ironically, hotspots constitute some of the best settings to study mantle/surface interactions because they are the direct result of matter and energy transfer from the mantle to the surface. The study of isostatic movements associated with hotspot activity may provide useful constraints on the physical mechanisms governing that interaction. Oceanic hotspots are particularly useful because the simpler architecture of oceanic plates allows a better separation

R. A. S. Ramalho, *Building the Cape Verde Islands*, Springer Theses,
DOI: 10.1007/978-3-642-19103-9_1, © Springer-Verlag Berlin Heidelberg 2011

between the several components that drive isostatic imbalance and because the thinner plate magnifies the isostatic effects acting on the lithosphere.

Oceanic hotspots disturb the isostatic balance in several ways. The more obvious mechanism, and probably the most widely studied, concerns the mass transfer to the surface caused by hotspot-driven volcanic activity. Ocean island volcanoes constitute some of the most prominent and rapidly-formed features on the Earth's surface. Consider, for example, the type locality for hotspot volcanism: the Big Island of Hawaii. When measured from the seafloor this island is the highest mountain on the planet, towering more than 10 km above the surrounding abyssal plain. Its volume (and weight) is such that the underlying plate is inflected downwards by more than 3 km and is still not in vertical equilibrium [1]. Despite the enormous size of the edifice, the Big Island is a relatively recent geological structure: it is probably less than 1 Ma old. Furthermore, the Big Island is only the most recent structure of a 6,000 km long chain—the Hawaiian-Emperor volcanic chain. This time-progressive linear array of volcanic islands and seamounts is the result of long-term hotspot activity (>80 Ma) on a fast-moving plate. The change in the edifices' characteristics along the chain has become the type example from which the classical view of island evolution was derived [3]. Likewise, the progressive loading of the Hawaiian chain became a primary example for isostatic studies on fast-moving plates [1].

The process of mass transfer to the surface associated with hotspot volcanism is not, in fact, completely efficient. Due to reasons still poorly understood, some of the rising material may reach a point of neutral buoyancy along the way and stall at crustal levels [1, 4]. The accumulation of this material will also play its role in the isostatic equilibrium, acting as an increasing subsurface load as it cools. The role of subsurface loads in modifying the isostatic balance associated with hotspot volcanoes has long been inferred to explain particular geometries of flexed lithosphere at hotspots [1, 4]. However, little is known about this mechanism and its evolution through time.

Another mechanism associated with hotspots that forces an isostatic feedback from the system concerns the generation of hotspot swells. Hotspot swells consist of broad regions of uplifted topography or shallower-than-expected bathymetry surrounding mid-plate volcanism. Swells constitute some of the largest bathymetric anomalies in the oceans, and are normally associated with positive gravity and geoid anomalies [5]. Their shape normally reflects the kinematic environment where they occur: it varies from an axisymmetric dome-shaped elevation on stationary plates to a wide parabolic shaped rise on fast-moving plates [6]. Hotspot swells are probably formed by buoyancy forces acting on or below the lithosphere. However, the source of this buoyancy is still a cause of hot debate. Current models normally involve hotspot-induced thermal rejuvenation [7], dynamic pressure from the ascending plume [6, 8] and ponding of a depleted harzburgitic swell root leftover from partial melting [5].

To understand the source of buoyancy that supports hotspot swells is a key goal in the comprehension of hotspot governing dynamics [9]. A limiting characteristic of most geophysical methods employed to study hotspot swells is that they provide a snapshot of current conditions and they are incapable of looking into the geological past. However, to understand what the mechanism is, or mechanisms are, behind swell formation, one needs to understand how the swell evolved through time and the relationship between that evolution and the history of hotspot volcanism. In other words, one needs to investigate the temporal component of the regional isostatic movements associated with swell development, and test its correlation with variations in the magma input. Similarly, the process of island building may hold key information about hotspot activity and swell formation.

Much of the research carried out with the objective of constraining swell dynamics is, inevitably, based on numerical models. Numerical modelling is a powerful tool that allows us to formulate possible solutions in the inverse approach to problem solving. As a result, several solutions and mechanisms have been proposed for swell development, mostly based on fluid dynamics, thermal expansivity and plate flexure (e.g., [5, 6, 10, 11]). However, physical evidence is needed to test the applicability of such models to reality. This physical evidence lies in the geological record.

The evolution of oceanic hotspot swells is reflected in their regional isostatic movements. Such vertical movements will have a direct impact on the sedimentation conditions on the seafloor they displace, and on the processes of island evolution taking place across the swell. Thus, the geological record may hold evidence regarding the isostatic changes that have occurred in the swell region. While seafloor stratigraphy potentially provides a direct correlation with swell development, its study is logistically difficult because it involves deep-sea drilling. The study of long-term island isostasy, on the other hand, provides a logistically easier approach to the problem. However, as previously outlined, other mechanisms may affect island isostasy and to understand the role of each mechanism in the process, one needs to decompose the resulting signal into its components.

Hence, the study of long-term isostatic movements affecting ocean island volcanoes yields insight into the processes associated with hotspot dynamics, via: (1) swell development; (2) surface loading associated with island development; (3) subsurface loading. Each of these components potentially has a distinct isostatic signature, trackable and quantifiable, and the relative role of each mechanism can thus be assessed. In addition, ocean island volcanoes are good case-studies in terms of isostasy because: (1) they are generated by a mass transfer that, albeit with some uncertainty and difficulties, that can be quantified both in volume and in time; (2) they lie on oceanic plates, which exhibit a simpler and thinner architecture when compared to continents; and (3) the geometry of the surrounding bathymetric surface is generally easier to interpret.

One way to reconstruct the history of isostatic movements affecting ocean islands is to study the changes occurring in island freeboard through the evolutionary history of the edifices, i.e., to look at the relative position of sea-level in each of the stages of edifice development and correlate it with the contemporaneous eustatic sea-level. The reconstruction of the islands' evolutionary history is thus essential to understand the changes in island freeboard. Additionally, it also provides information regarding the magmatic history of the ocean island volcanoes, the latter parameter being essential for freeboard/volcanism correlations.

The Hawaiian-Emperor island/seamount chain has become the prototype for hotspot-generated oceanic islands [3]. For this reason, the Hawaiian chain probably constitutes the most intensively studied group of oceanic hotspot volcanoes. However, the Hawaiian chain is the product of a hotspot on a fast-moving plate. In this kind of setting, the movement of the plate away from the melting source offers an obvious mechanism to extinguish volcanism, and the linear array of islands only allows along-chain comparisons. In contrast, in a stationary plate environment, volcanism remains stationary with respect to the melting source and the effects are cumulative. The stationary condition may also allow comparisons between edifices located at the same distance from the swell centre, because islands normally occur in a cluster instead of in a linear chain. Such stationary conditions can be found in the Cape Verde Archipelago, a group of islands derived from an oceanic hotspot situated offshore from Western Africa.

This thesis reports on research undertaken to reconstruct the history of vertical movements affecting the Cape Verde Islands and goes on to test the likely mechanisms that caused those movements and to investigate their link with swell forming processes.

1.2 The Growth and Decay of Ocean Island Volcanoes

The growth and decay of oceanic hotspot volcanoes is intrinsically related to a competition between volcanic construction and erosive destruction. Ultimately, since erosion never ceases to act, it is the variation in the rate of magma-supply through geological time that controls the edifice's evolution. In detail, however, other factors may influence the evolutionary history of oceanic volcanoes. These are: (1) the nature, intensity, style of activity and timing of the volcanism; (2) the nature and intensity of the erosive agents that act upon the edifices; (3) the relative position of sea-level; and (4) the role of biological colonisation in the sedimentary processes affecting the edifices.

1.2.1 The Prototypical Hawaiian Example

The evolutionary stages that characterise the growth and decay of ocean island volcanoes were initially defined through the Hawaiian example [3, 12–18]. Hawaiian volcanoes, as currently perceived, evolve through seven stages: submarine pre-shield stage, shield-building stage, capping stage, erosional stage, rejuvenated stage, atoll stage, and guyot stage [17].

The submarine pre-shield stage is also called the deep marine stage or initial stage. Volcanic activity normally starts with the intrusion of feeder dykes and sills into pelagic sediments, producing nonexplosive peperites and inducing intrusion-related uplift of the ocean bottom [19]. When volcanism disrupts through sediment, a cone of pillow and sheet lavas starts to form. Extrusion normally occurs under hydrostatic pressure so effusive eruptions are dominant. Eruptions rates are normally low and typically involve low volumes of alkalic basalt [13–15, 17–19]. Intrusions are frequent and account for a large portion of the edifice. As the seamount grows, a steep-sided edifice is created, commonly featuring a defined caldera and two or more rift zones radiating from the summit [13, 15, 17, 19, 20]. The steep profile is normally maintained by mass wasting processes acting on the unstable pillow and hyaloclastite slopes, contributing to the development of a volcanic apron [19]. As time progresses, eruptions become stronger and more frequent and the edifice enters the shield-building stage.

The shield-building stage is characterised by increasing eruption rates involving large volumes of tholeiitic basalts. This stage probably accounts for >95% of the volume of each volcano and usually lasts between 0.5 and 1 Ma [3, 13, 15, 17, 18]. Three substages, however, have been recognised: the submarine substage, the sea-level substage, and the subaerial substage. The submarine substage evolves from the pre-shield stage through an increase in eruption rates and volume of erupted materials. Simultaneously, magmas normally pass from alkalic to tholeiitic. The edifice rapidly grows and starts to assume a moderately sloping aspect. The increasing surface load causes subsidence and bending of the lithosphere underneath [1, 21, 22]. As the seamount grows upwards, increasingly lower hydrostatic pressures allow the onset of vesicularity and explosive volcanism starts to occur. When the edifice approaches the surface, the volatile pressure overcomes hydrostatic pressure and explosive phreatomagmatic eruptions predominate. The sea-level or explosive substage is thus reached. During this substage, surtseyan eruptions create submarine tuff cones and large quantities of hydroclastic material that get redeposited on the flanks of the edifice and in the volcanic apron [19]. The upward growth of tuff cones above sea-level marks the transition to the subaerial substage [13, 15, 18, 19]. However, due to the predominance of unconsolidated products and the proximity to sea-level, the edifice is particularly vulnerable to wave erosion and so morphologies are normally ephemeral. Only if volcanism exceeds erosion, is a more permanent island created. This is generally achieved when the vent area gets isolated from the sea and effusive eruptions build a resistant lava cap that allows the stabilisation of the island edifice [19]. The eruptive style thus changes from surtseyan (explosive) to hawaiian/strombolian (effusive) and the subaerial substage starts. A small shield-like volcano starts to evolve by the upward accumulation of successive effusive flows and by flank growth achieved by the generation of lava deltas on the fringes of the edifice. Since aa and especially pahoehoe lavas become dominant, flank growth and the advancement of coastlines is further enhanced by tube-feeding processes, extending the distances the flows can reach before cooling [17]. The ongoing activity eventually builds up the gently sloping convex-up shield volcano of enormous proportions that characterise young Hawaiian islands (e.g. Mauna Loa).

Intrusions also account for a significant fraction of the total volume of shield volcanoes [15, 17, 23]. Since the eruptive activity is normally concentrated along rift zones, large dyke complexes are formed beneath these areas [23]. The intrusion of these complexes is normally accommodated by lateral slip on deep faults. These may coincide with the contact between the base of the volcano and the underlying seafloor sediments [17, 20, 23], and may be the locus of landslides [24, 25]. In fact, large mass-wasting events are another important aspect of large shield volcanoes, especially during the shield-building stage when the volcanoes are close to their maximum size [17].

The capping stage gets its name from the irregular carapace of low-silica alkali basalts that caps the edifice after the shield-building stage. This stage is normally short-lived, marks a drastic reduction in magma-supply rate and a transition back to highly alkalic basalts [13, 15, 17, 18]. Eruption style is consequently more explosive and the formation of strombolian cinder cones is favoured, leading to a more irregular morphology. Likewise, the piling of more pasty lavas near the source vents of the summit region or along rift zones steepen the slopes of the volcano [17]. Not all Hawaiian volcanoes experience this stage, however.

As soon as volcanism stops, erosion assumes a dominant role and starts the destruction of the edifice. The volcano then enters into the erosional stage. Marine abrasion and stream erosion incise the volcano, generating high coastal cliffs and deep canyons that erase the original constructional morphology. As the island subsides, fringing coral reefs grow and the process of atoll formation starts. Many volcanoes, however, experience a period of renewed volcanism after 3–5 Ma of quiescence [3, 13, 15, 17, 18].

The rejuvenated stage consists of a short period of renewed volcanism, with eruptions producing small volumes of highly alkali lavas [12, 13, 15, 17]. The generation of monogenetic cones is favoured, producing cinder cones inland and surtseyan tuff rings in coastal areas. Lavas generated inland frequently infill the deeply incised topography and flow through valleys, creating lava deltas when they reach the sea. This stage is also called post-erosional stage.

Following the rejuvenated stage, volcanoes experience a long period of erosion and subsidence that eventually razes them to sea-level [26]. During this period many islands become surrounded by a peripheral coral reef [26–29]. With the end of active volcanism and the continuing erosion and subsidence, the edifice eventually disappears beneath the waves [13, 17]. At the same time the reef grows upwards, keeping pace with the subsidence, and forms a ring-shaped structure termed an atoll [27, 28]. The atoll stage is thus reached. Eventually, with continuous subsidence, even atolls drown and form typical flat-topped seamounts called guyots. This stage is called the gyuot or drowned island stage.

1.2.2 The Representativeness of the Hawaiian Example

The evolutionary stages defined through the Hawaiian example do not constitute a uniformly applicable model for ocean island development. Despite some basic similarities with Hawaii, such as the dominance of basaltic shields, many ocean island volcanoes seem to have developed differently [3]. The differences found across archipelagos are normally the result of contrasting melting rates, chemical compositions of magma, and the different geodynamic setting in which archipelagos occur. Some of the most striking differences are discussed below.

The transition from alkalic to tholeiitic basalts observed in the Hawaiian islands during the shield-building stage is rare in other settings. In most oceanic islands the shield-building stage is normally composed of alkalic basalts. In fact, the transition to tholeiitic seems to be restricted to settings with high degrees of partial melting [3]. This aspect has obvious implications. As a result of the lower magma-supply rates and more alkalic lavas, the dominant subaerial eruptive style is normally strombolian as opposed to hawaiian. Consequently, pyroclastic deposits are normally more abundant and lava flows reach lower extents than in Hawaiian volcanoes, resulting in smaller and steeper edifices. Steeper composite edifices may also imply higher rates of flank collapse [3]. Another implication concerns the submarine stages of island evolution. Since in most oceanic volcanoes there is not a steep increase in magma-supply rates, or a change from alkalic magmas to tholeiitic, the distinction between pre-shield stage and shield stage is less clear. For this reason, some authors e.g. [19] prefer a different classification for the early evolutionary stages of ocean islands that distinguishes between seamount stage (with deep water and intermediate water substages), emergent island stage, and island subaerial shield-building stage.

Another difference concerns the lifetime of individual islands. Atlantic islands seem to last longer than Hawaiian islands [3, 25]. For example, while in Hawaii the subaerial shield-building stage normally lasts 0.5 Ma, in the Canaries the same stage may last up to an order of magnitude longer [25]. In a similar fashion, quiescent periods are also longer, and several post-erosional phases may occur [3, 30].

Atolls are structures whose occurrence is restricted to particular conditions of subsidence and to oceanographic conditions favourable to coral growth. As a consequence many old islands do not become atolls. These normally get slowly truncated by wave and stream erosion, leading to a low-relief morphology with extensive peripheral wave-cut surfaces. The long-term erosion together with the absence of uplift will eventually lead to the full destruction of the exposed edifice, generating a flat-topped guyot. This type of guyot, however, is different in its genesis because is not formed by the drowning of a reef structure but by the truncation of the volcanic edifice; the formation of a classic guyot is dominated by subsidence while the latter is dominated by erosional processes.

Due to these differences, and due to the lack of a unifying model in the literature, it becomes necessary to define a more general model for ocean island evolution. Since many of the observations on which the present work is based are of island evolution, it is important to define, at this stage, the terms and concepts used later in this manuscript. Here, a modification of the classical "Hawaiian" model will be used, comprising eight stages as follows.

1. *Seamount stage* The seamount stage comprises the edifice's phases of development in an entirely submarine environment, i.e., from the initiation of

volcanism to the growth of the edifice in shallow waters. This stage may be divided into a deep-water substage and an intermediate water/shoaling substage; the first occurs below the critical depth for explosive volcanism and normally comprises exclusively effusive products, while the second is characterised by the onset of increasing vesicularity of lavas and hydroclastic and/or pyroclastic explosive activity [14, 15, 17–19].

2. *Emergent island stage* This stage corresponds to the transition from the submarine to the subaerial environment and is characterised by surtseyan activity. Large volcaniclastic deposits are normally produced during this stage due to the combination of a surtseyan eruptive style and marine erosion, transport and re-sedimentation processes. Structures are frequently ephemeral, so many islands do not exhibit direct evidence from this period of edifice growth.
3. *Shield-building (subaerial) stage* This stage starts when the vents are isolated from the sea and the eruptive style passes from surtseyan to strombolian/hawaiian. The subaerial period of edifice growth thus starts, being dominated by effusive activity. During this stage flank growth is characterised by the formation of lava delta sequences on the fringes of the edifice, so large volumes of submarine lavas may be produced but from onshore vents.
4. *Capping stage* This stage is only definable if an abrupt change in lava composition and/or eruptive rates occur. Thus, it may be indistinct from the last period of shield-building and, unless those changes are observed, it may be considered as part as the shield-building stage.
5. *Erosional stage* This stage starts with the end of the shield-building or capping stages and is characterised by volcanic quiescence and erosion. During this stage the destruction of the edifice intensifies and the morphology gradually changes from constructional to erosive landforms.
6. *Rejuvenated stage* An edifice is in the rejuvenated (or post-erosional) stage when a period of renewed volcanism follows erosion. This stage frequently creates valley-filling sequences and coastal lava deltas, and normally involves low volumes of alkalic magma. On many islands, several periods of post-erosional volcanism may occur, sometimes separated by long periods of quiescence.
7. *Atoll/Razed island stage* This stage is characterised by the absence of volcanic activity and the slow destruction of the reminiscent edifice. If subsidence and coral growth occur, atoll structures are created. In the absence of subsidence and coral growth, the island is slowly truncated by stream and marine erosion, forming a razed morphology.
8. *Guyot (or drowned island) stage* This stage is characterised by the submersion of the island edifice, either due to the outpacing of subsidence relative to coral growth, or due to the truncation of the edifice by marine erosion.

1.2.3 Island Evolution on Fast Moving vs Stationary Plates

The above outlined differences in island evolution are probably the indirect result of contrasts in plate velocity, age and thickness, and melting rate. We may, consequently, identify some of the expected differences between archipelagos formed on fast moving plates and archipelagos formed on stationary plates. For example, the longer life-time of individual islands and the existence of several post-erosional volcanic periods on archipelagos situated on a stationary plate are probably caused by the relative fixity between the hotspot and the melting source. On fast moving plates, the movement of the plate away from the melting source offers an obvious mechanism for the waning of volcanism, while on stationary plates other mechanisms must be responsible for the decay of volcanic activity, perhaps lithospheric thickening. On stationary plates, any decay of hotspot swell height is probably much slower than on fast-moving plates, where the movement of the plate will have an effect on the buoyancy mechanism supporting the swell. The characterisation of these differences is key to understanding the processes behind hotspot activity and island building.

1.3 Aims and Workflow

The research associated with the current thesis was carried out with the following aims:

- To reconstruct the history of vertical movements affecting the Cape Verde Archipelago, the type-example of an archipelago formed by hotspot activity on a stationary plate environment.
- To infer the cause/effect relationship between those movements and island evolution.
- To identify the different mechanisms that drive island freeboard in a stationary plate environment and their relative role in modifying the isostatic equilibrium of ocean islands in such settings.
- To understand the hotspot processes that might be behind the changes in island freeboard, namely the phenomenon of swell development and its possible driving mechanisms.

In order to achieve the proposed goals, our research was based in a simple workflow which is also reflected in the structure of this thesis.

In Chap. 2 the Cape Verde Archipelago and its main characteristics are introduced. The general geography and geology of the archipelago are presented here for reference, as well as the geodynamic context in which the Cape Verdes occur.

In Chap. 3 the expected contribution of the regional uplift associated with swell development is computed and discussed. A fluid dynamical numerical model is tested in order to predict the possible uplift/subsidence pattern associated with Cape Verde swell evolution. Likewise, the shape of the possible swell root underlying the Cape Verde swell is also computed. The material present in this chapter will be further developed in the future, and a manuscript will be submitted to a specialised journal.

Chapter 4 introduces the methodology used to trace island freeboard. It discusses the possible reference levels that are usable for island freeboard reconstructions, and explains why sea-level was preferred over other reference data. The caveats of using sea-level as a reference level are presented, and the choice of a particular eustatic curve over another is discussed. The concept of vertical displacement is introduced and a general overview of the main palaeo-markers of sea-level found on ocean islands is presented, with details regarding the estimation of past sea-level positions.

In Chap. 5 an overview of the main tracers of uplift and subsidence found on each of the Cape Verde islands is presented. The volcanostratigraphy of each island is analysed in detail and the information about relative sea-level change through the different stages of island evolution is extracted, based on the information available in the literature and new field observations. Preliminary island freeboard reconstructions are also attempted, and the possible causes of uplift are briefly discussed. An abbreviated version of the Chap. 5 is published in the "Journal of the Geological Society".

Chapter 6 concerns the dating of a set of palaeo-markers of sea-level from the Cape Verdes. A number of strategic palaeo-markers of sea-level was chosen from the available range described in Chap. 5 and dated. The geochronological methods used in this work—strontium isotope stratigraphy, Ar/Ar and U-Th—are introduced here and the methodological details are also presented. The resulting numerical ages are presented and their validity is discussed.

Chapter 7 focuses on the reconstruction of island freeboard changes through geological time. It uses the vertical movement reconstructions of two particular islands—Santiago and São Nicolau—and, together with the information from the rest of the archipelago, discusses the possible causes for the significant differential island uplift inferred for the Cape Verdes. It tests several possible mechanisms and models and infers the causes behind island uplift. This chapter has been submitted for publication in the journal "Marine Geology" and published.

Finally, Chap. 8 presents the conclusions. It compiles a summary of the individual conclusions from each chapter and enumerates the effects of island freeboard in the construction of the Cape Verde Islands.

References

1. Watts, A. (2001). *Isostasy and flexure of the lithosphere*. Cambridge, UK: Cambridge University Press.
2. Maddox, J. (1998). *What remains to be discovered: Mapping the secrets of the universe, the origins of life, and the future of the human race*. New York, USA: Martin Kessler Books.
3. Schmincke, H.-U. (2004). *Volcanism* (1st ed.). Berlin: Springer Verlag.
4. Watts, A., & Ten Brink, U. (1989). Crustal structure, flexure, and subsidence history of the Hawaiian Islands. *Journal of Geophysical Research (Solid Earth), 94*(B8), 10473–10500.

5. Morgan, J., Morgan, W., & Price, E. (1995). Hotspot melting generates both hotspot volcanism and a hotspot swell. *Journal of Geophysical Research (Solid Earth), 100*(B5), 8045–8062.
6. Sleep, N. (1990). Hotspots and mantle plumes: Some phenomenology. *Journal of Geophysical Research (Solid Earth), 95*(B5), 6715–6736.
7. Detrick, R., & Crough, T. (1978). Island subsidence, hot spots, and lithospheric thinning. *Journal of Geophysical Research (Solid Earth), 83*(B3), 1236–1244.
8. Ribe, N., & Christensen, U. (1999). The dynamical origin of Hawaiian volcanism. *Earth and Planetary Science Letters, 171*(4), 517–531.
9. Lodge, A. (2006). Seismic constraints on swell formation beneath the Cape Verde Islands (Ph.D. thesis, University of Bristol).
10. Olson, P. (1990). Hot spots, swells and mantle plumes. In: M. Ryan (Ed.), *Magma transport and storage* (chap. 3, pp. 33–51). Chichester, UK: Wiley.
11. Ribe, N., & Christensen, U. (1994). Three-dimensional modeling of plume-lithosphere interaction. *Journal of Geophysical Research (Solid Earth), 99*(B1), 669–682.
12. Stearns, H. (1940). Four-phase volcanism in Hawaii. *Geological Society of America, 51*, 1947–1948.
13. Macdonald, G., Abbott, A., & Peterson, F. (1983). *Volcanoes in the sea: the geology of Hawaii* (2nd ed.). USA: University of Hawaii Press.
14. Clague, D., & Dalrymple, G. (1987). *The Hawaiian-Emperor volcanic chain. Part I. Geologic evolution in volcanism in Hawaii* (vol. 1, pp. 5–54). U.S. Geological Survey Professional Paper.
15. Clague, D., & Dalrymple, G. (1989). Tectonics, geochronology, and origin of the Hawaiian-Emperor volcanic chain. In E. Winterer, D. Hussong, & R. Decker (Eds.), *The geology of North America, Volume N, The Eastern Pacific Ocean and Hawaii* (vol. N, pp. 188–217). USA: Geological Society of America.
16. Moore, J., & Clague, D. (1992). Volcano growth and evolution of the island of Hawaii. *Geological Society of America Bulletin, 104*(11), 1471–1484.
17. Orton, G. (1996). Volcanic environments. In H. Reading (Ed.), *Sedimentary environments: Processes, facies and stratigraphy* (pp. 485–567). Oxford, UK: Blackwell.
18. Clague, D., & Dixon, J. (2000). Extrinsic controls on the evolution of Hawaiian ocean island volcanoes. *Geochemistry, Geophysics, Geosystems, 1*(4), 1–9.
19. Schmidt, R., & Schmincke, H.-U. (2000). Seamounts and island building. In H. Sigurdsson, B. Houghton, S. McNutt, H. Rymer, & J. Stix (Eds.), *Encyclopedia of volcanoes* (pp. 383–402). USA: Academic Press.
20. Walker, G. (1999). Volcanic rift zones and their intrusion swarms. *Journal of Volcanology and Geothermal Research, 94*(1–4), 21–34.
21. Moore, J. (1970). Relationship between subsidence and volcanic load, Hawaii. *Bulletin of Volcanology, 34*(2), 562–576.
22. Walcott, R. (1970). Flexure of the lithosphere at Hawaii. *Tectonophysics, 9*(5), 435–446.
23. Walker, G. (1992). Coherent intrusion complexes in large basaltic volcanoes—a new structural model. *Journal of Volcanology and Geothermal Research, 50*(1–2), 41–54.
24. Moore, J., Clague, D., Holcomb, R., Lipman, P., Normark, W., & Torresan, M. (1989). Prodigious submarine landslides on the Hawaiian Ridge. *Journal of Geophysical Research (Solid Earth), 94*(17), 17465–17484.
25. Carracedo, J. (1999). Growth, structure, instability and collapse of Canarian volcanoes and comparisons with Hawaiian volcanoes. *Journal of Volcanology and Geothermal Research, 94*(1–4), 1–19.
26. Menard, H., & Ladd, H. (1963). Oceanic islands, seamounts, guyots and atolls. *The Sea, 3*, 365–385.
27. Darwin, C. (1842). *On the structure and distribution of coral reefs*. London, UK: Ward Lock.
28. Grigg, R. (1982). Darwin point: A threshold for atoll formation. *Coral Reefs, 1*(1), 29–34.

29. Scott, G., & Rotondo, G. (1983). A model to explain the differences between Pacific plate island-atoll types. *Coral Reefs, 1*(3), 139–150.
30. Geldmacher, J., Hoernle, K., van den Bogaard, P., Zankl, G., & Garbe-Schönberg, D. (2001). Earlier history of the 70-Ma-old Canary hotspot based on the temporal and geochemical evolution of the Selvagen Archipelago and neighboring seamounts in the eastern North Atlantic. *Journal of Volcanology and Geothermal Research, 111*(1–4), 55–87.

Chapter 2
The Cape Verde Archipelago

2.1 Introduction

The Cape Verde Archipelago, also known as the Cape Verde Islands or simply the Cape Verdes, is a group of islands situated in the North Atlantic approximately at latitudes between 14°40′ and 17°30′ N and longitudes between 21°30′ and 25°30′ W of Greenwich (see Fig. 2.1). The islands lie 450–600 km off the western coast of Africa at the approximate latitude of the Cape Verde promontory in Senegal, from which they take their name.

The archipelago was discovered in 1460 AD by Portuguese sailors exploring the west coast of Africa. Human settlement started soon after but suffered numerous setbacks from the hardships imposed by the arid climate. Now, the archipelago comprises the Republic of Cape Verde. Nine of the ten islands are inhabited, hosting a population of $\sim$450,000 inhabitants.

The origins of the Cape Verde Archipelago are attributed to the long-term mid-plate volcanism associated with the Cape Verde hotspot. The volcanic activity probably started in the Oligocene/Miocene and extended well into the Holocene. Historical eruptions are unknown except on Fogo, a highly active volcano whose last eruption occurred in 1995.

2.2 Geography of the Archipelago

2.2.1 Geomorphology

The archipelago is composed of ten islands and eight minor islets arrayed in a west-facing horseshoe. The islands are traditionally divided into the Barlavento (windward) group, comprising the islands of Santo Antão, São Vicente, Santa Luzia, São Nicolau, Sal, and Boa Vista, and the Sotavento (leeward) group, comprising Maio, Santiago, Fogo, and Brava. However, a close look at the

R. A. S. Ramalho, *Building the Cape Verde Islands*, Springer Theses,
DOI: 10.1007/978-3-642-19103-9_2, © Springer-Verlag Berlin Heidelberg 2011

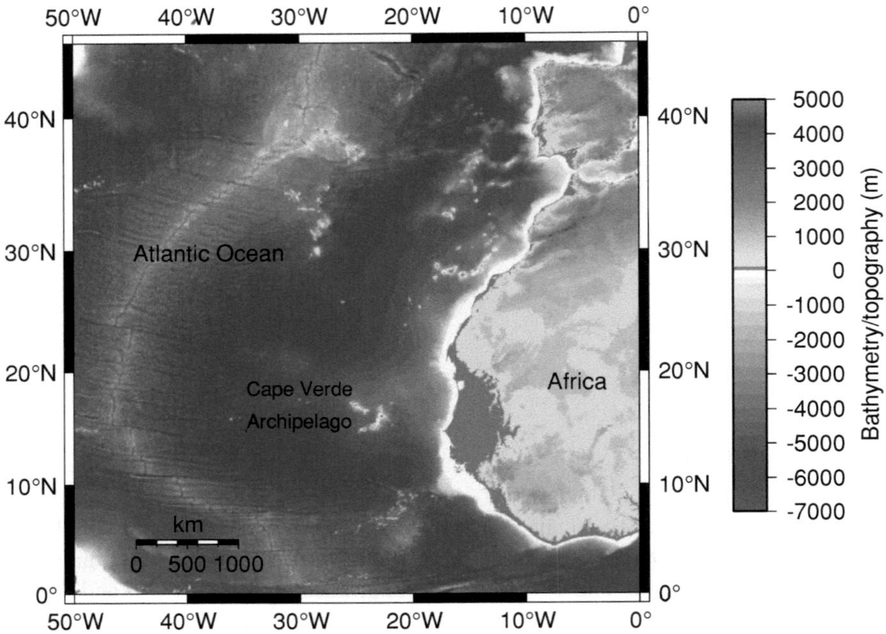

Fig. 2.1 Location of the Cape Verde Archipelago. Elevation data from Amante and Eakins [1]

regional bathymetry suggests a different division based on the coalescence of the submarine pedestal of the islands' edifices: it seems that the archipelago is composed of a "northern chain", comprising the islands of Santo Antão, São Vicente, Santa Luzia, and São Nicolau, and a "east-to-southern chain", comprising Sal, Boa Vista, Maio, Santiago, Fogo, and Brava (see Fig. 2.2). For simplicity, hereafter we will refer to the latter chain only as the "southern chain". Fogo and Brava form a partly detached edifice, because their pedestals coalesce with Santiago slightly above the 3,000 m isobath, which is significantly deeper than elsewhere. We consider Fogo and Brava as part of the southern chain, however.

In the southern chain, the bathymetry reveals the presence of a shallow area between Boa Vista and Maio, known locally as the João Valente Bank or João Valente Shoals. This feature, whose flat summit is just 14 m below the present sea-level, probably corresponds to a former island whose top has been truncated by marine erosion, i.e., it is probably a guyot. In the northern chain, the islands of São Vicente, Santa Luzia, and the Islets of Ilhéu Raso and Ilhéu Branco (to the ESE of Santa Luzia) probably constitute a single elongated edifice, as suggested by the bathymetry. These were probably "separated" by recent sea-level rise, as attested by the shallow channels existing between these islands, which barely reach 20 m depth.

The bathymetry also reveals that several seamounts are present in the vicinity of the main islands. Most of these seamounts correspond to volcanic edifices that never reached sea-level, since flat topped morphologies seem to be rare. Among the

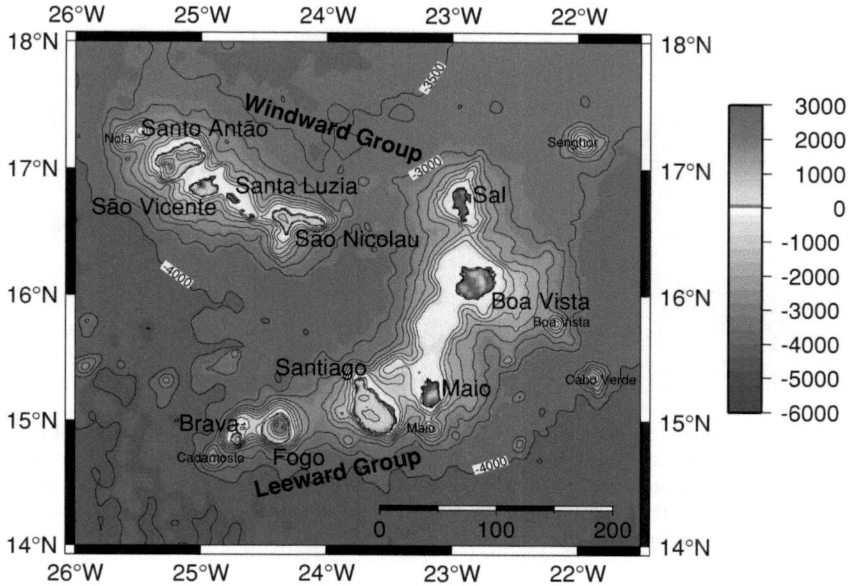

Fig. 2.2 Relief of the Cape Verde Archipelago. Contour lines are in m. Note the contrast in relief between the eastern and western islands. The western islands tend to exhibit higher reliefs than the eastern ones, denoting younger landscapes in the former and denuded topographies in the latter. Note also the north–south asymmetry in the insular platforms of the islands, with more extensive platforms in the northern part of the edifices. Elevation data from Amante and Eakins [1]

Table 2.1 Cape Verde area by island (in km^2)

Southern chain from E to W						Northern chain from E to W			
Sal	B. Vista	Maio	Santiago	Fogo	Brava	S. Nicolau	S. Luzia	S. Vicente	S. Antão
216	620	269	991	476	67	388	35	227	779

most important seamounts lying in the vicinity of the Cape Verde Islands we highlight: the Senghor Seamount ($\sim$100 km to ENE of Sal), the Boa Vista and Cabo Verde Seamounts (ESE of Boa Vista), the Maio Seamount (S of Maio), the Cadamosto Seamount (WSW of Brava), and the Nola Seamount (WNW of Santo Antão)—see Fig. 2.2.

The archipelago's total land area corresponds to about 4,068 km^2 unevenly distributed among the islands. The largest island (Santiago) is 991 km^2 as opposed to the just 35 km^2 covered by the smallest island (Santa Luzia) (see Table 2.1). Inter-island distances vary from just 8 km (between São Vicente and Santa Luzia) to 270 km (between Santo Antão and Maio).

There is a contrast in relief between the eastern and western islands (see Figs. 2.2, 2.3 and Table 2.2). The eastern islands of Sal, Boa Vista and Maio are low and flat-lying and their maximum elevations are under 450 m above sea-level (asl).

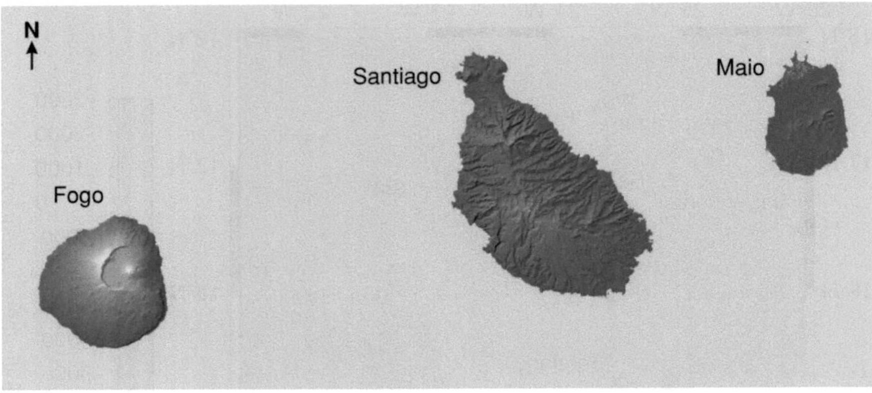

Fig. 2.3 Morphological types of edifices. Note the contrast in the surface morphology of these three edifices. Maio (in the E) is the oldest edifice with a flat lying erosive morphology. Santiago represents an intermediate stage, where the morphology of the (shield) edifice is still perceptible, despite being cut by deep valleys. Fogo (in the W) is a young stratovolcano with a preserved morphology, only interrupted by the east-facing caldera-collapse scar

Table 2.2 Islands' maximum elevation distribution (in m asl)

Southern chain from E to W						Northern chain from E to W			
Sal	B. Vista	Maio	Santiago	Fogo	Brava	S. Nicolau	S. Luzia	S. Vicente	S. Antão
406	387	436	1,394	2,829	976	1,340	395	725	1,979

Their general landscape is old and denuded, and erosive landforms prevail. Their only relief is normally a few isolated residual morphologies, normally erosion-resistant phonolitic edifices, or the cones formed during the last post-erosional eruptive stage. These constitute razed edifices whose relief has been levelled mostly by periods of partial immersion, where wave-cut erosion dominated. The morphology denotes the late post-erosional stage. Conversely, the western islands generally exhibit higher relief and reach elevations that frequently surpass 1,000 m asl. Their landscape is young and the primary volcanic forms are still perceptible. Edifices in an early post-erosional stage, like Santiago and Santo Antão, still exhibit the overall shield morphology despite being cut by deep valleys. The still active volcano of Fogo, with an elevation of 2,829 m asl and the highest point in the archipelago, corresponds to a stratovolcano whose morphology is only interrupted by the east-facing caldera-collapse scar, now partially masked by the construction of the new summit cone. This volcano is still in its main building stage and erosive features are rare.

2.2.2 Climate and Erosional Regime

The Cape Verdes enjoy a tropical oceanic climate. Annual temperature variation is small and ranges from 19 to 29°C. However, since the archipelago is part of the

Sahelian arid belt, rainfall levels are lower than in other West African countries. This condition imposes an arid to semi-arid regime with scarce but concentrated rain during a "wet season" from August to October. Annual precipitation barely reaches 260 mm/a and is also unevenly distributed due to orographic effects. Rainfall is normally confined to short but heavy downpours associated with tropical storms, inducing a torrential regime to fluvial erosion. These episodic but strongly erosive flow conditions, when combined with the prominent topography of younger volcanic landforms, lead to a deeply incised drainage pattern characterised by deep canyons and narrow gorges. It is indeed remarkable how these ephemeral streams, which are probably active erosive agents for only a minute fraction of the average year, are capable of such erosive power. The deep canyons testify to their powerful effects [2]. In addition, the arid conditions also favour overland flow and soil wash which, together with slope mass wasting, contribute to the high load transported by the drainage system and lead to increased erosive power. However, as soon as the stream leaves the steep and narrow flow channel, the erosive and transport capability of the stream wanes, leading to the formation of large alluvial fans when the geomorphic conditions are propitious. In denuded islands, and due to the lack of relief, drainage often assumes a different pattern. Drainage, in these areas, is normally characterised by braided systems and incipient meanders with low lying channels. During storms, riverine flow frequently bursts the channel banks and mantles large areas with the entrained sediments it carries. This, together with high overland flow and soil wash, leads to a barren and dusty landscape covered in thin layers of outwash deposits whose finer elements are easily remobilised by the wind.

The islands are subjected to a normally persistent northeasterly wind, inducing an energetic marine erosion regime with asymmetrical characteristics. Coastlines facing the prevailing winds are thus more energetic and endure higher erosion rates. This leads to an asymmetrical erosive topography with higher sea cliffs and more extensive insular platforms (below sea-level) in the northern slopes of the volcanic edifices (see Fig. 2.2). The more rapid coastline retreat in the northern shores also influences the drainage base level, leading to steeper stream profiles and increasing the erosive power of stream flow on the northern slopes.

2.3 Geodynamical Setting

The Cape Verde Islands lie on the African tectonic plate, a plate that has moved very little in the last 30 Ma in relation to the hotspot reference frame [3, 4]. Moreover, the pole of rotation for the African plate seems to be located in the vicinity of the Cape Verde Archipelago, further contributing to the fixity of the archipelago relatively to the underlying hotspot [5]. This might explain why the Cape Verde Islands are arranged as a cluster rather than a single linear chain. For whatever reason, we may consider the archipelago stationary relative to its melting source [6–8]. This situation enables the study of plume/plate interactions without the complication of plate movement [7].

Despite its proximity to the African mainland, the Cape Verde Archipelago clearly lies on oceanic lithosphere. This is directly confirmed by the presence of underlying Mesozoic ocean floor at the nearby IODP sites [9, 10]. It is indirectly confirmed by the presence of seafloor spreading magnetic anomalies mapped across the region surrounding the islands, with the Cretaceous Magnetic Quiet Zone lying eastwards of these islands [11, 12]. According to Williams et al. [12], the Neogene volcanic episode creating the Cape Verde Archipelago did not mask the original texture and fracture zones of the Mesozoic seafloor, attesting to fact that these islands lie on oceanic crust. Seismic investigations also suggest an underlying oceanic crustal and mantle structure [8, 13, 14]. The underlying lithosphere is interpreted to be between 120 and 140 Ma old, as suggested by the presence of the Mesozoic seafloor spreading magnetic anomalies M0–M16 [11, 12, 15] (see Figs. 2.4 and 2.5). Thus, if no thermal rejuvenation occurred, the islands lie over thick and cold lithosphere that has reached its asymptotic properties [16]. In fact, it seems that the features of the original Mesozoic seafloor in the nearby area of the islands are still discernible, suggesting that no regional reheating occurred [12].

The disposition of the Cape Verde Islands along two apparently different trends, one NW–SW and other NE–SW, is suggestive of a possible relationship with fracture zones [12, 13, 17]. Mata et al. [18] suggested that the linear arrangement

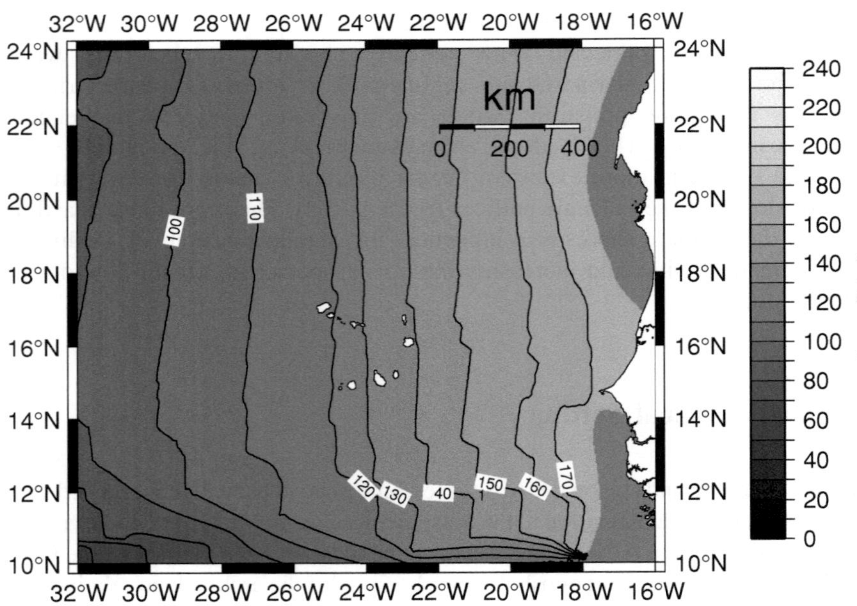

Fig. 2.4 Age of the lithosphere in the vicinity of the Cape Verde Archipelago. Note that the islands lie on 120–140 Ma old lithosphere. Data from Muller et al. [15]

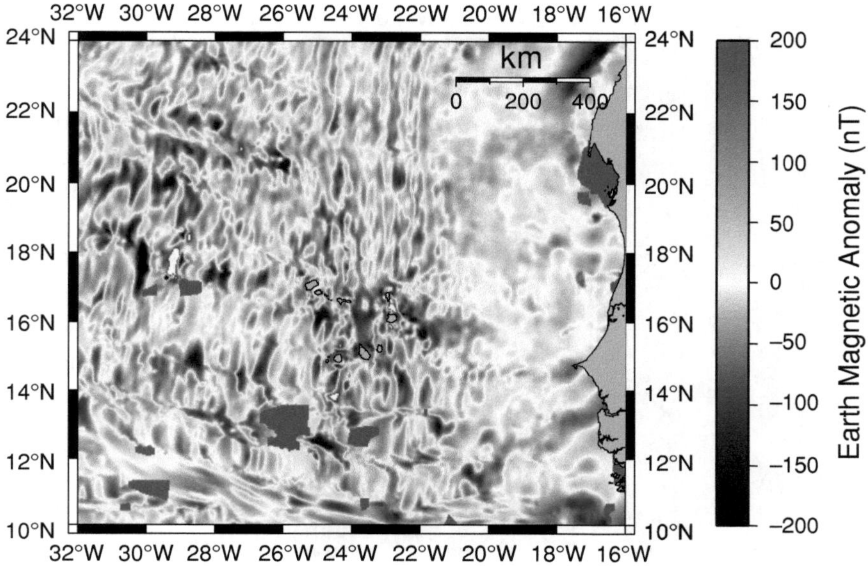

Fig. 2.5 Seafloor spreading magnetic anomalies in the vicinity of the Cape Verde Archipelago. Data from Maus et al. [19]

of the Cape Verde Islands in two distinct chains may result from the interaction between the hotspot and deep fractures. However, this geometry may be fortuitous since there is no clear evidence for the islands being located preferentially along fracture zones [12].

2.4 The Cape Verde Rise and Islands

The Cape Verde Islands rest upon one of the most prominent features in the NE Atlantic: the Cape Verde Rise [9]. The Cape Verde Rise is a large smooth dome-shaped swell, spanning almost 10° in latitude and reaching depths as shallow as ~3,000 m [9] (Fig. 2.6). When compared to the expected depth for the age, the Cape Verde Rise stands in excess of 2,000 m above the reference value, making it the largest bathymetric anomaly in the world's oceans [6, 20–22]. The origins of the Cape Verde Rise are still controversial, but it is widely accepted that this feature is associated with the Cape Verde hotspot volcanism [6, 20, 21]. The stratigraphic record inferred from the existing IODP sites strongly suggests that the rise is the result of a broad uplift in excess of 2,000 m that occurred sometime around the Oligocene/Miocene transition or the Early Miocene [9, 10].

The volcanic activity associated with the Cape Verde hotspot probably started sometime in the Late Oligocene or Early Miocene. This seems to be the age of the oldest exposed lavas in the archipelago that are unequivocally associated with hotspot activity [23]. Likewise, this is also the suggested age for the formation of

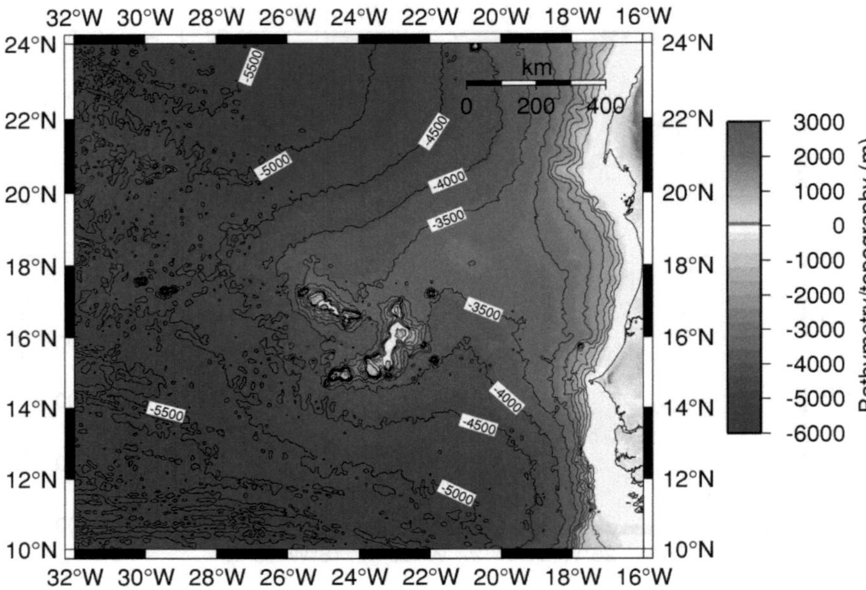

Fig. 2.6 Map of the Cape Verde Rise. Contour lines in meters. Elevation data from Amante and Eakins [1]

the Cape Verde Rise, as inferred by the sedimentary record offshore the Cape Verdes [9, 10]. However, the activity was not continuous and not all the islands were formed at the same time. It seems that Sal, Boa Vista and Maio were formed first, as attested by their surface geology and published geochronology (see Fig. 2.7). Around 10–8 Ma ago, volcanism seems to have started further west, creating the basement complexes of Santiago, São Nicolau and Santo Antão. Soon after a vigorous period of activity resumed, leading to the voluminous shield-building stage of Santiago and possibly the entire northern chain. Sal experienced post-erosional volcanic activity as well, and possibly the basement complexes of Fogo and Brava were created by that time. This vigorous Miocene/Pliocene period of activity is well documented in the sediments of the Cape Verde Rise [9, 10]. The bulk of the archipelago, consequently, seems to have been formed towards the end of the Neogene [24]. Holocene volcanism has been reported on several islands, but historical volcanism is not known except on Fogo. The highly active volcanic centre of Pico do Fogo has erupted every 20 years, on average, since the discovery of the archipelago [25, 26].

Oceanic intra-plate volcanism is normally dominated by low-viscosity magmas with relatively low levels of entrapped volatiles. As a consequence, effusive activity normally prevails, leading to the generation of broad shield-type edifices. These edifices may assume an elongated or star-shaped morphology due to occurrence of fissure eruptions along preferential directions or rift arms, but the main factors controlling these geometries are still poorly understood. More towards the end of

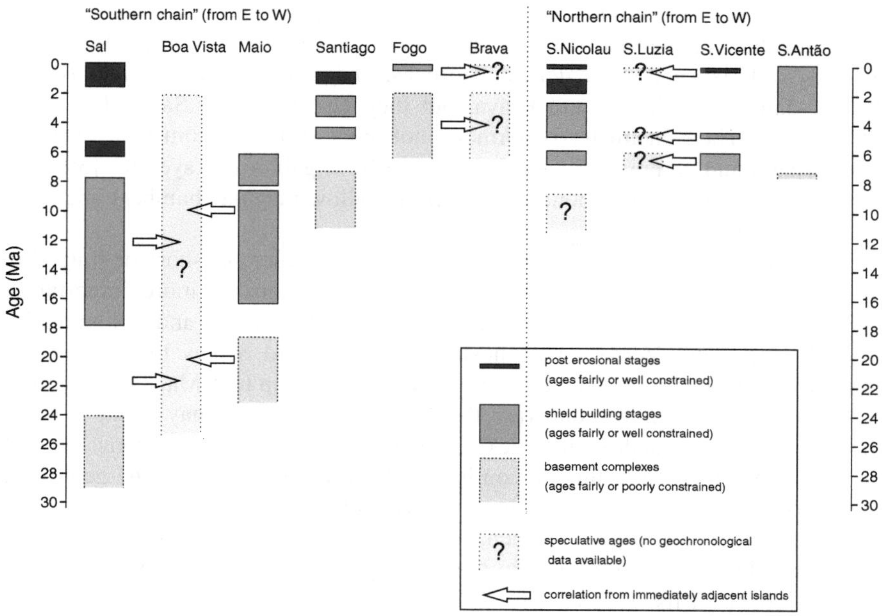

Fig. 2.7 Age distribution of exposed volcanic/intrusive products across the Cape Verde Archipelago. Modified from Holm et al. [24] and based on the geochonological data of Bernard-Griffiths et al. [27], Mitchell et al. [28], Torres et al. [23], Plesner et al. [29], Jørgensen and Holm [30], Duprat et al. [31], Holm et al. [24], Foeken et al. [32]. Note the possible age progression of the oldest exposed rocks in the southern chain and the approximately synchronous extrusion of all the islands composing the northern chain

each eruptive cycle, the eruption of more evolved magmas may sporadically occur, leading to the formation of stratovolcanoes. However, these are normally rare.

The petrological array of lavas extruded in the vast majority of oceanic islands is normally dominated by alkali basalts and their derived products. The Cape Verdes are no exception and the overwhelming majority of their lavas correspond to under-saturated lithologies; silica-saturated lithologies are extremely rare [2]. However, what sets apart the Cape Verde magmatism from those of the other Atlantic archipelagos is the presence of intrusive and extrusive carbonatites (normally calcio-carbonatites) in at least five of the ten islands that compose the archipelago [30, 33]. The carbonatites, once thought to be confined to the basement complexes of the islands, actually occur in the mature stages of some of the islands [33].

According to Kogarko [34], the vast majority of the Cape Verde lavas can be classified in two main differentiated series: (1) a high-alkali series comprising picrites, high-Mg foidites, low-Mg foidites and phonolites; (2) a moderate alkalinity series comprising picrobasalts, basanites, tephrites, tephrophonolites, phonotephrites, phonolites and trachytes. The lavas are thus dominantly relatively primitive lavas, generated by low degrees of partial melting at considerable depths [7, 35].

The presence of plutonic cores within the basement complexes of the edifices is also a notable characteristic of the Cape Verdes. Unsurprisingly, these lithologies are only well exposed in the denuded edifices. They are better represented on Boa Vista, Maio, São Vicente and Brava, but they also occur on Sal and Santiago [33, 36–42]. Their chemistry is normally not much different from their extrusive counterparts and comprise lithologies ranging from gabbros to syenites [39]. The plutonic bodies may correspond to the top of shallow magma chambers unearthed by deep levels of erosion.

Volcanic activity is not a continuous process, neither on short or long time scales. During the periods between individual eruptions or, more importantly, during the longer-term periods of volcanic quiescence, erosion and sedimentation prevail. The Cape Verde's volcanostratigraphic record shows that quiescence usually lasts from 1 to 4 Ma, but occasionally may last up to 6 Ma. This condition seems to be normal in the context of the Atlantic islands, and may be the result of the hotspot's fixity and/or the thick lithosphere above it. The outcome of such periods of erosion and sedimentation is usually well represented in the geological record of the islands, either by massive deposits of lahar or extensive marine carbonates. It is not surprising, then, that many islands exhibit considerable volumes of sedimentary formations in their stratigraphy, or thin but extensive sedimentary cover on its surface.

Another defining feature of Cape Verde is the significant inter-island differences in the surface geology across the archipelago. These differences mostly reflect the different ages of the edifices and/or the level of erosion experienced by the islands. However, differences may also arise from rather different structural controls on the volcanism (fissure vs central eruptive style) and different uplift/subsidence histories.

A complete overview on the volcanostratigraphy of each island is presented on Chap. 5, with the available age constraints. Simplified geological maps are also presented in that chapter.

2.5 A Heterogeneous Mantle Source

The partial melting occurring in the mantle beneath oceanic islands is perceived to be the result of adiabatic decompression of plume and asthenosphere components in the ascending flow [43]. Thus, ocean island lavas, and particularly the more primitive basalts, are thought to generally reflect the geochemical properties of the source and the physical conditions prevailing during melting [43]. Since plumes are considered the most effective probe of mantle reservoirs, ocean island basalts have been extensively studied for the last three decades in order to provide insight on mantle chemical properties [44].

From the advent of the hotspot theory, the origin of the Cape Verde Islands is believed to involve volcanic activity associated with an underlying mantle plume [45]. The Cape Verdes are thus viewed as the surface manifestation of a process that taps a deep mantle reservoir [18, 24, 45–48]. As in many other hotspots, though, the mantle source associated with the Cape Verde volcanism seems to be heterogeneous [7, 24, 29, 30, 45–48]. In fact, the Cape Verde has been suggested to be an extreme example of a zoned mantle plume [7, 24].

A general overview of the petrology and the variation of basaltic rocks seems to confirm that these are quite similar on the different islands [45, 46]. The suite of lavas that characterise the bulk of the Cape Verdean volcanism seems to reflect parental magmas that correspond to high-Mg foidites, picrobasalts and basanites [34], indicating a low degree of partial melting of mantle peridotite at considerable depth [35]. The occurrence of carbonatites on many islands of the archipelago additionally suggests that the mantle source is CO_2-rich [7]. However, despite this apparently common source, the variation found in the isotopic compositions of lavas from different islands requires that the magmatism had different source regions on different islands [30, 45, 46].

A characteristic of the Cape Verde magmatism is the occurrence of significant intra-island and inter-island radiogenic isotope variations, reflecting temporal and spatial changes in the mantle source [7, 47]. The heterogeneity of the Cape Verde mantle source was first shown by Gerlach et al. [45], using Sr–Nd–Pb isotopes. These authors highlighted the differences between the Southern and Northern Islands, and interpreted them as a mix of three isotopically distinct end-members: (1) Depleted MORB Mantle (DMM); (2) a HIMU-like end-member; and (3) an EM1-like end-member [47]. According to Gerlach et al. [45], the northern islands reflect a strong HIMU (high time-integrated U/Pb) component, possibly supplemented by trapped DMM material, while the southern islands reflect an EM1-like signature possibly due to mixing.

More recently, Doucelance et al. [47] suggested that the intra-island variability in the Pb–Sr–He isotope composition found in Cape Verde requires at least five distinct end-members for the source of the archipelago's basalts. According to these authors, the northern islands reflect a signature resulting from recycled oceanic crust and long term storage, with variable amounts of entrained mixed material. In contrast, the southern islands, though also requiring the presence of recycled oceanic crust and long term storage, additionally have an isotopic signature that is strongly influenced by subcontinental lithospheric mantle, yielding EM1-like trace element patterns [47]. São Nicolau is somehow slightly different from the other northern islands, and requires the assimilation of unaltered Jurassic MORB [47]. However, it has been shown that the oceanic lithosphere plays an important role in controlling the geochemical signature of ocean island basalts, altering their original major and trace elements compositions via its age/thickness, and suggesting that shallow-level interactions need to be taken into account [44].

The isotopic variations seem not to be confined to the basaltic lithologies. Jørgensen and Holm [30] suggested the existence of two distinct HIMU sources within the Cape Verde plume, in order to explain the contrasting isotopic signatures between basalts and carbonatites. According to Jørgensen and Holm [30], the isotopic signatures found in the lavas of São Vicente suggest an evolution from a deep-seated plume-dominated source to a significantly depleted mantle source at higher levels. Carbonatite contamination of younger basaltic magmas is also proposed by Jørgensen and Holm [30]. Noble gas isotopic compositions of carbonatites also suggest a relatively undegassed source, possibly from the lower mantle [48].

2.6 Summary

The Cape Verde Islands constitute an archipelago generated by Neogene oceanic hotspot volcanism, standing on the broad bathymetric anomaly known as the Cape Verde Rise. The archipelago's stationary position with respect to the melting source, makes it an ideal place to study plume/plate interactions without the complexities imposed by plate movement. The nature of the volcanism is essentially effusive, corresponding to alkaline and silica-undersaturated lavas generated by partial melting at deep levels, and associated with a heterogeneous mantle source. The islands exhibit significant differences in their surface geology, mostly due to differences in the age, erosion level and uplift/subsidence histories experienced by the edifices.

References

1. Amante, C., & Eakins, B. W. (2009). ETOPO1 1 arc-minute global relief model: Procedures, data sources and analysis. *NOAA Technical Memorandum NESDIS NGDC-24*, 19.
2. Mitchell-Thomé, R. (1972). Outline of the geology of the Cape Verde Archipelago. *International Journal of Earth Sciences,61*(3), 1087–1109.
3. Burke, K., & Wilson, J. (1972). Is the African Plate stationary? *Nature,239*(5372), 387–390.
4. Morgan, W. J. (1983). Hotspot tracks and the early rifting of the Atlantic. *Tectonophysics,94*(1–4), 123–139.
5. Pollitz, F. (1991). Two-stage model of African absolute motion during the last 30 million years. *Tectonophysics,194*(1–2), 91–106.
6. McNutt, M. (1988). Thermal and mechanical properties of the Cape Verde Rise. *Journal of Geophysical Research (Solid Earth),93*(B4), 2784–2794.
7. Holm, P., Wilson, J. R., Christensen, B., Hansen, L., Hansen, S., Khein, K. M., et al. (2006). Sampling the Cape Verde plume: Evolution of melt compositions on Santo Antão, Cape Verde Islands. *Journal of Petrology,47*(1), 145–189.
8. Lodge, A., & Helffrich, G. (2006). Depleted swell root beneath the Cape Verde Islands. *Geology,34*(6), 449–452.
9. Lancelot, Y., Seibold, E., Cepek, P., Dean, W., Eremeev, V., Gardner, J., et al. (1978). Site 368: Cape Verde Rise. In Y. Lancelot, E. Seibold, P. Cepek, W. Dean, V. Eremeev, J. Gardner, et al. (Eds.), *Initial reports of deep sea drilling project* (vol. 41, pp. 233–326). Washington: U.S. Government Printing Office.
10. Faugeres, J. C., Legigan, P., Maillet, N., & Latouche, C. (1989). Pelagic, turbiditic, and contouritic sequential deposits on the Cape Verde plateau (leg 108, site 659, Northwest Africa): sediment record during Neogene time. In *Proceedings of the Ocean Drilling Program, Scientific Results* (vol. 108, pp. 311–328).
11. Hayes, D., & Rabinowitz, P. (1975). Mesozoic magnetic lineations and the magnetic quiet zone off northwest Africa. *Earth and Planetary Science Letters,28*(2), 105–115.

12. Williams, C., Hill, I., Young, R., & White, R. (1990). Fracture zones across the Cape Verde Rise, NE Atlantic. *Journal of the Geological Society,147*(5), 851–857.
13. Dash, B., Ball, M., King, G., Butler, L., & Rona, P. (1976). Geophysical investigation of the Cape Verde archipelago. *Journal of Geophysical Research (Solid Earth),81*(29), 5249–5259.
14. Pim, J., Peirce, C., Watts, A., Grevemeyer, I., & Krabbenhoeft, A. (2008). Crustal structure and origin of the Cape Verde Rise. *Earth and Planetary Science Letters,272*(1–2), 422–428.
15. Muller, R., Sdrolias, M., & Roest, W. (2008). Age, spreading rates and spreading symmetry of the world's ocean crust. *Geochemistry Geophysics Geosystems,9*(Q04006), 1525-2027.
16. Stein, C., & Stein, S. (1992). A model for the global variation in oceanic depth and heat flow with lithospheric age. *Nature,359*(6391), 123–129.
17. Bebiano, J. (1932). A geologia do Arquipélago de Cabo Verde. *Comunicações dos Serviços Geológicos de Portugal,18*, 167–187.
18. Mata, J., Torres, P., Munhá, J., Mendes, M., & Silva, L. (1999). Tectonic controls on the geochemistry of Sal Island (Cape Verde). *European Union of Geosciences,*10, *Journal of Conference,* 4(1), 348.
19. Maus, S., Barckhausen, U., Workgroup, B., Muller, D., & Fairhead, D. (2008). EMAG2: A 2-arc-minute resolution global magnetic anomaly grid compiled from satellite, airborne and marine magnetic data. In *American Geophysical Union,* Fall Meeting 2008 (Abstract# GP53A-0776).
20. Crough, S. (1978). Thermal origin of mid-plate hot-spot swells. *Geophysical Journal International,55*(2), 451–469.
21. Sleep, N. (1990). Hotspots and mantle plumes: Some phenomenology. *Journal of Geophysical Research (Solid Earth),95*(B5), 6715–6736.
22. Monnereau, M., & Cazenave, A. (1990). Depth and geoid anomalies over oceanic hotspot swells: A global survey. *Journal of Geophysical Research (Solid Earth),95*(B10), 15–429.
23. Torres, P., Silva, L., Serralheiro, A., Mendes, M., Macedo, J., & Gomes, A. (2002). Geologia da Ilha do Sal. Comunicações do Instituto de Investigação Científica Tropical.
24. Holm, P., Grandvuinet, T., Friis, J., Wilson, J. R., Barker, A. K., & Plesner, S. (2008). An 40Ar-39Ar study of the Cape Verde hot spot: Temporal evolution in a semistationary plate environment. *Journal of Geophysical Research (Solid Earth),113*(B8), B08201.
25. Torres, A., & Soares, J. (1946). Formações Sedimentares do Arquipélago de Cabo Verde. I—Actualização de conhecimentos. Junta das Missões Geográficas e de Investigações Coloniais.
26. Torres, P., Madeira, J., Silva, L., Brum da Silveira, A., Serralheiro, A., & Mota Gomes, A. (1995). Carta geológica das erupções históricas da Ilha do Fogo: revisão e actualização.
27. Bernard-Griffiths, J., Cantagrel, J.-M., Alves, C., Mendes, F., Serralheiro, A., & Macedo, J. (1975). Geochronologie: Donnés radiometriques potassium-argon sur quelques formations magmatiques des îlles de l'archipel du Cap Vert CR. *Seances Academy of Sciences Series D,280*, 2429–2432.
28. Mitchell, J., Bas, M. L., Zielonka, J., & Furnes, H. (1983) On dating the magmatism of Maio, Cape Verde Islands. *Earth and Planetary Science Letters,64*(1), 61–76.
29. Plesner, S., Holm, P. M., & Wilson, J. R. (2002). 40Ar-39Ar geochronology of Santo Antão, Cape Verde Islands. *Journal of Volcanology and Geothermal Research,120*(1–2), 103–121.
30. Jørgensen, J., & Holm, P. (2002). Temporal variation and carbonatite contamination in primitive ocean island volcanics from São Vicente, Cape Verde Islands. *Chemical Geology,192*(3–4), 249–267.
31. Duprat, H., Friis, J., Holm, P., Grandvuinet, T., & Sørensen, R. (2007). The volcanic and geochemical development of São Nicolau, Cape Verde Islands: Constraints from field and 40Ar/39Ar evidence. *Journal of Volcanology and Geothermal Research,162*(1–2), 1–19.
32. Foeken, J., Day, S., & Stuart, F. (2009). Cosmogenic 3He exposure dating of the Quaternary basalts from Fogo, Cape Verdes: Implications for rift zone and magmatic reorganisation. *Quaternary Geochronology,4*(1), 37–49.
33. Mourão, C., Mata, J., Doucelance, J., Madeira, J., Brum da Silveira, A., Silva, L., & Moreira, M. (2010). Quaternary extrusive calciocarbonatite volcanism on Brava Island (Cape Verde). *Journal of African Earth Sciences,56*(2–3), 59–74.

34. Kogarko, L. (2008). Characteristics of alkali magma differentiation at the Cape Verde Islands. *Geochemistry International,46*(11), 1071–1080.
35. Paepe, P. D., Klerkx, J., Hertogen, J., & Plinke, P. (1974). Oceanic tholeiites on the Cape Verde Islands: Petrochemical and geochemical evidence. *Earth and Planetary Science Letters,22*(4), 347–354.
36. Serralheiro, A. (1966). Contribuição para o conhecimento geológico da Ilha de S. Vicente (Cabo Verde). *Garcia de Orta, Serviços Geológicos,14*(1), 139–152.
37. Serralheiro, A. (1970). Geologia da Ilha de Maio (Cabo Verde), Junta de Investigações do Ultramar.
38. Serralheiro, A., Alves, C., Macedo, J., & Silva, L. (1974). Note preliminaire sur la géologie de l'île de Boa Vista (Cap-Vert). *Garcia de Orta. Serviços Geológicos,1*(3), 53–60.
39. Serralheiro, A. (1976). A Geologia da Ilha de Santiago (Cabo Verde). *Boletim do Museu e Laboratorio Mineralógico e Geológico da Faculdade de Ciências,14*, 157–369.
40. Stillman, C., Furnes, H., Le Bas, M., Robertson, A., & Zielonka, J. (1982). The geological history of Maio, Cape Verde Islands. *Journal of the Geological Society,139*(3), 347–361.
41. Silva, L., Serralheiro, A., Macedo, J., Gomes, A., & Torres, P. (1990). Carta Geológica de Cabo Verde, Ilha do Sal, na escala de 1/25000 (folhas 1-2), Edição do Instituto Investigação Científica Tropical/Instituto de Cooperação Económica.
42. Torres, P., Silva, L., Serralheiro, A., Tassinari, C., & Munhá, J. (2002). Enquadramento geocronológico pelo método K/Ar das principais sequências vulcano-estratigráficas da Ilha do Sal—Cabo Verde. *Garcia de Orta, Serviços Geológicos,18*(1–2), 9–13.
43. Wilson, M. (1989). *Igneous petrogenesis.* Berlin: Springer.
44. Millet, M.-A., Doucelance, R., Schiano, P., David, K., & Bosq, C. (2008). Mantle plume heterogeneity versus shallow-level interactions: A case study, the São Nicolau Island, Cape Verde archipelago. *Journal of Volcanology and Geothermal Research,176*(2), 265–276.
45. Gerlach, D., Cliff, R., Davies, G., Norry, M., & Hodgson, N. (1988). Magma sources of the Cape Verde Archipelago: Isotopic and trace element constraints. *Geochimica et Cosmochimica Acta,52*, 2979–2992.
46. Davies, G., Norry, M., Gerlach, D., & Cliff, R. (1989). A combined chemical and Pb-Sr-Nd isotope study of the Azores and Cape Verde hot spots; the geodynamic implications. *Geological Society of London Special Publications, 42*(1), 231–255.
47. Doucelance, R., Escrig, S., Moreira, M., Gariépy, C., & Kurz, M. D. (2003). Pb-Sr-He isotope and trace element geochemistry of the Cape Verde Archipelago. *Geochimica et Cosmochimica Acta,67*(19), 3717–3733.
48. Mata, J. (2006). Noble gases isotopic signatures from Cape Verde oceanic carbonatites. *Geophysical Research Abstracts,8*, 07462.

Chapter 3
Constraining the Cape Verde Swell Using Numerical Models

3.1 Introduction

Hotspot swells are defining features of mid-plate volcanism and consist of broad areas surrounding the volcanic centre where topography is anomalously high [1, 2]. These features are particularly evident in the oceans, where crustal structure is governed by heat transfer [3, 4]. Oceanic hotspot swell dimensions can range from 600 to 1,600 km in width and reach up to $\sim$2 km high [5], making them some of the most prominent features in ocean basins. Another striking characteristic of mid-plate swells is their association with broad positive geoid anomalies that are nearly linearly related to topography [1]. Due to these outstanding features, and because a key to understanding hotspots lies in understanding the origins of their associated swells [6], many researchers over the last 40 years have focused their research on hotspot swells (e.g. [1, 2, 5, 7–9]).

The persistence of swells means that they are in isostatic balance. An upwards force—provided by a buoyancy source—is necessary to counteract the pull of gravity in order to sustain swells through time [10]. The swell's shape and size is mostly governed by three parameters: (1) the characteristics of the buoyancy source; (2) the characteristics of the overlying "rigid" shell; and (3) the movement of the plate. Necessarily, a few fundamental questions arise. What is the source of buoyancy? What is the interaction between the source of buoyancy and the overlying shell? How does the system evolve through time? The answer to these questions are still intensely debated. There are three proposed mechanisms for hot spot swell development: (1) thermal rejuvenation, which relates isostatic uplift with thermal erosion and thinning of the lithosphere induced by the hotspot enhanced heat flux [1, 11], a similar situation to the ponding of hot, low-density plume material [7, 12]; (2) dynamic pressure, which associates the uplift of the swell with the upward flux associated with the plume, which spreads laterally below the plate [7, 13, 14]; (3) compositional buoyancy associated with the ponding at the base of the plate of low-density depleted material leftover from melting and melt extraction [2]. These models are incomplete descriptions of all hotspot phenomena.

R. A. S. Ramalho, *Building the Cape Verde Islands*, Springer Theses,
DOI: 10.1007/978-3-642-19103-9_3,

Thermal rejuvenation explains well the rapid uplift and subsequent slow decay of the swell due to passage of the lithosphere over the hotspot [10], and the apparent lithospheric thinning that seems to occur in some hotspots [15]. However, it fails to explain how the hotspot is able to rapidly thin the lithosphere to the necessary extent to fully create the swell [5, 10, 16, 17], and in many hotspots there is no evidence for significant lithospheric thinning [5, 7–9]. Lithospheric thinning would also create a measurable heat anomaly that, so far, has proved elusive to find [10], and difficult to measure [18]. The swell decay away from the melting source also seems to be faster than expected by a fully thermally driven model [2], suggesting that heat loss is not the only, or even main, process behind swell decay [2, 14]. The dynamic pressure model, on the other hand, is consistent with the shape of some swells [7, 10] and also predicts the rapid uplift and subsequent slow decay [10]. However this model implies a spreading material flowing away from the hotspot (made of plume material) and that feature, which if hot would generate slower-than-average seismic velocities even away from the centre of the swell, seems to be absent at least in the Cape Verde case [8]. Dynamic pressure also offers a poor match to the observed geoid anomalies [10, 17]. Furthermore, since the dynamical model is associated with asthenospheric rather than lithospheric compensation, very high geoid to topography ratios (>6 m/km) are expected and these seldom occur [19]. Finally, the depleted swell root model explains the rapid uplift and subsequent decay, as well as the apparent volume relationship between the swell and the island edifices [2]. This model is also compatible with the faster-than-average seismic velocities observed in the Cape Verde swell mantle structure because it is the only model that allows for a "cool" root for the swell [2, 8]. None of the models, however, seems to offer a consistent explanation for the post-shield volcanism away from the hotspot centre. Whether all may satisfy the observational constraints presented by different hotspots and their geodynamical settings seems to add additional complexity to the problem. Many of these mechanisms seem to act together [2, 7, 16, 20], making the driving process elusive.

Since "direct" observations are mostly limited to the topography, geoid, heat flow and the crustal/upper mantle seismic structure of swells, the use of numerical models is thus essential to test the multitude of viable solutions that satisfy the observational constraints. This chapter presents an attempt to numerically model the shape of the Cape Verde swell to yield insight on swell formation processes.

3.2 The Cape Verde's Bathymetric Swell

The Cape Verde Rise is the largest bathymetric anomaly in the oceans [11]. It is considered a hotspot swell associated with the mid-plate volcanism responsible for the building of the Cape Verde archipelago, whose history spans $\sim$26 Ma [21]. The Cape Verde hotspot occupies a unique position among major hotspots due to its proximity to the pole of rotation of the African plate, a semi-stationary plate [2, 5, 7, 16, 22], thus making it stationary with respect to its melting source [22–24].

The Cape Verde swell dimensions is 1,200–1,600 km in width, and its crest reaches 1.9–2.3 km above the surrounding seafloor [5, 16, 25] (Fig. 3.1). The flat bulge that constitutes the top of the swell is connected to the African continental rise by a broad sill of slightly deeper waters [26]. This morphology is probably not related to the swell itself, but formed by the accumulation of sediments between the African margin and the top of the swell [11], as suggested by the eastwards increase in sediment thickness. This feature makes it difficult to constrain the eastern edge of the swell. However, if a sediment correction is performed it confirms that the eastwards projecting sill is a sediment artifact. Despite the difficulties in precisely defining its dimensions, the morphology of the swell seems elongated east-west (see Fig. 3.1) and its base roughly coincides with the 5,400 m bathymetric curve (the expected depth for 120–140 Ma old lithosphere). The islands themselves are clustered slightly off-centre to the southwest, where they seem to be disposed in two chains forming a rough semi-arc [16].

Sediment thicknesses along the Cape Verde Rise increase from west to east [28], reaching about 1,000 m in the highest area of the rise [27, 29, 30], and thicknesses well in excess of 4,000 m closer to the African margin. The sedimentary pile comprises a lower package, with increasing thickness towards the east and composed of Upper Jurassic to Lower Miocene limestones, shales and turbidites, and an upper package of post Lower Miocene calcareous oozes and volcaniclastic sediments associated with the Cape Verde's volcanic apron [27–30]. In the eastern part of the rise, a wedge of terrigenous sediments of Saharan origin greatly thickens the package. The fact that very few turbidites have reached the top of the Cape Verde Rise in the last 25 Ma, together with the lack of volcaniclastic

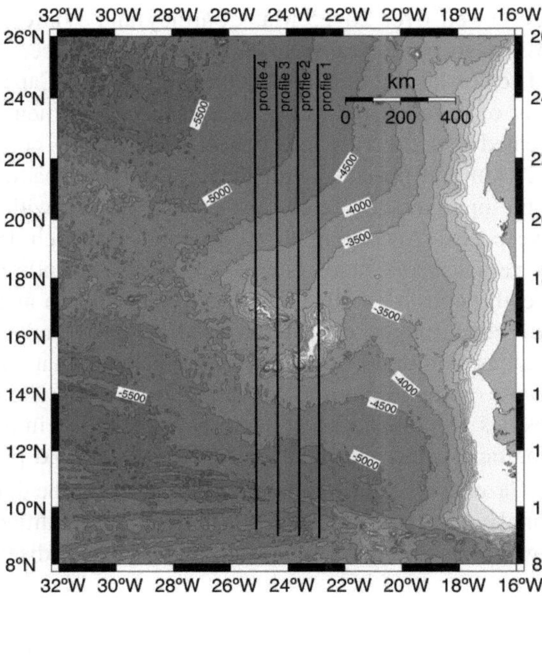

Fig. 3.1 Bathymetry of the Cape Verde Swell, and location of bathymetric profiles. Contour lines in meters. Elevation data from Amante and Eakins [26]

material in the sedimentary pile prior to that age, suggests that the uplift of the Cape Verde Swell must have occurred sometime around the Oligocene/Miocene passage [30]. This age is in rough agreement with the oldest exposed lavas in the Archipelago [21]. In addition, in the vicinity of the archipelago, the flexural moat seems to be tilted outwards from the islands, suggesting a broad regional uplift that partially counteracts the effects of island loading [28].

Classified in the literature as a "thermal" swell [19], the Cape Verde Rise coincides with a positive geoid anomaly between 8 and 12 m [5, 16, 25, 31], leading to one of the highest known geoid to topography ratios (3.4–6.0 m/km), a value well above the average trend [5, 19]. Since this geoid to topography ratio is dependent of the average depth of compensation [19], a low density root of altered lithosphere has been proposed as the cause of uplift associated with swell formation [2, 5, 6, 8, 16], with compensation depth values ranging from 40 km [25] to 69 km [16] or even 100 km [28]. Since the Cape Verde swell is considered to be fixed relative to its melting source and is overlain by old and thick lithosphere, it is not surprising that extreme values of geoid to topography ratios and uplift occur [5].

Courtney and White [31] analysed the heat flow across the Cape Verde Rise and found that this parameter increases towards the top, to values reaching 16 ± 4 mW m^{-2} above the background value of 45.5 ± 3.4 mW m^{-2}. Recently, Harris and McNutt [18] demonstrated that reliable heat flow measurements are very difficult to obtain because they are strongly dependent on fluid flow through the heterogeneous sedimentary cover, and measurements close to the surface seldom represent the real heat flow. Notwithstanding these caveats, the Courtney and White [31] thermal models showed that the observed heat flow is significantly below the expected value for a >20 Ma thermal resetting, and lithospheric reheating alone cannot explain the sustained uplift and volcanism on the Cape Verde swell over 20–25 Ma. Courtney and White [31] instead propose dynamic pressure to be the cause of the swell's existence.

The crustal and mantle structure beneath the Cape Verde swell was recently the focus of several studies. A teleseismic experiment by Lodge and Helffrich [8] revealed that the structure beneath the islands comprises a thickened crust, up to 22 km and possibly with local subsurface loading, underlain by a 60–80 km ($\sim$65 on average) thick low-density layer, with faster-than-average seismic wavespeeds. According to Lodge and Helffrich [8], the depths at which the low-velocity zone starts (suggesting normal mantle material) vary from $\sim$60 km beneath Fogo to $\sim$110 km beneath Sal, with an average depth around $\sim$80 km. This value can be considered as a lower bound for the swell root [8], and a compensation depth for the isostatic balance that supports the Cape Verde swell because it represents the transition from a low-density layer to normal mantle densities. A compensation depth at $\sim$80 km is in agreement with the 76–82 km theoretical value expected for a 120–140 Ma old plate (according to Monnereau and Cazenave [32] the compensation depth can be defined as $7t^{1/2}$). These features strongly suggest the existence of a thick but modified lithosphere—with chemical buoyancy as the source of uplift—and is in stark disagreement with enhanced plate thinning or underplating by warm material, as one would expect if thermal erosion or dynamic support were the driving forces for swell uplift. The crustal thickening, and especially the thickness of the low-density layer, increases eastwards suggesting the older islands to be underlain by greater volumes of underplated material [6, 8]. Similarly, the depths at which the low-velocity zone starts seem to be inversely correlated with the distance to the central area of the swell and directly correlated with the local swell height (see Fig. 3.2). This asymmetry seems to suggest that the low-density material is produced by hotspot melting and spreads from individual melting loci beneath the islands [6, 8]. More recently, Pim et al. [9] examined the crustal structure along a transect in the vicinity of Cape Verde using seismic refraction. Their study revealed what seems to be a normal crustal thickness for the age ($\sim$7 km), perhaps suggesting that the crustal

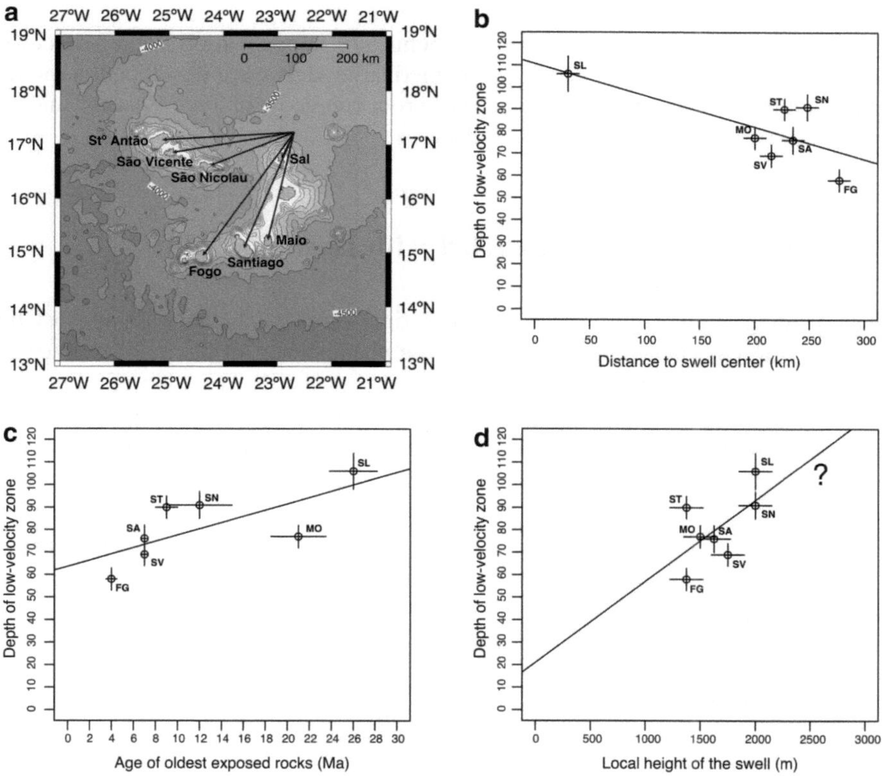

Fig. 3.2 Variation in the depth of the low velocity zone inferred by Lodge and Helffrich [8] with distance to centre of the swell, age of oldest exposed rocks and local swell height. **a** Location of the islands relative to the central area of the swell. **b** Depth of low-velocity zone vs distance to the swell central area. Note the possible inverse correlation between these two parameters. **c** Depth of low-velocity zone vs age of oldest exposed rocks. Note the possible direct correlation between these two parameters. **d** Depth of low-velocity zone vs local swell height (inferred from the bathymetry). Note the possible direct correlation between these two parameters

thickening proposed by Lodge and Helffrich [8] is confined to the islands' proximal area. Another interesting feature proposed by Pim et al. [9] is the anomalous low density mantle beneath the centre of the swell, with values about 30–105 kg/m^3 lower than normal mantle densities (3,300 kg/m^3).

Thus, in summary, we can define the Cape Verde swell as a broad topographic feature with 600–800 km radius, 1.9–2.3 km high and a super-gaussian shape [2, 28]. The swell seems to have been formed around 25 Ma ago, the same approximate age attributed to the earliest volcanism responsible for island building. The swell coincides with a positive geoid anomaly of 8–12 m, generating a high geoid to topography ratio and suggesting a compensation depth $\sim$80 km. The compensation depth is in general agreement with the crustal and mantle structure inferred from seismic wavespeeds, which seems to be characterised by a locally thickened crust (from 7 up to 22 km), underlain by a $\sim$65 km thick low-density, cool root—a chemically buoyant root.

To further test the hypothesis of a chemically buoyant root as the source of uplift of the Cape Verde's swell, two numerical models were developed: a model based on the expansion of an axisymmetric root coupled with pure isostasy, and an elastic flexural subsurface loading model.

3.3 Lubrication Model Coupled with Isostatic Uplift

The ponding of buoyant material beneath a rigid surface, coupled with an isostatic model, is a reasonable approximation to the phenomenon of plume/plate interaction and consequent swell formation [2, 6]. The geometric evolution of a buoyant viscous fluid, as an axisymmetric gravity current (Fig. 3.3), approximates the conditions arising after partial melting of peridotite under a stationary plate.

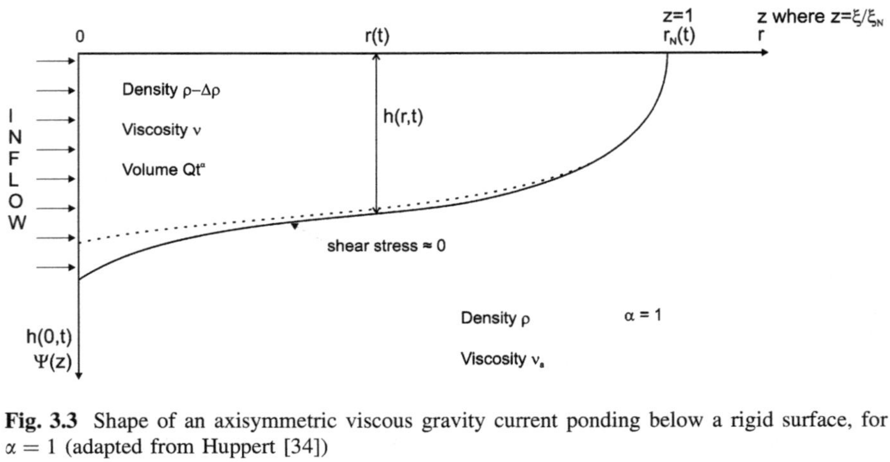

Fig. 3.3 Shape of an axisymmetric viscous gravity current ponding below a rigid surface, for $\alpha = 1$ (adapted from Huppert [34])

The preferential iron partitioning during melting produces a relatively iron-rich magma erupted to the surface to produce ocean island volcanism and leaves behind a refractory magnesium-rich residue [33]. Hence, melting affects the buoyancy distribution, not only by absorbing latent heat (reducing the thermal buoyancy), but mostly by generating chemical buoyancy, since the unmelted residue is less dense than the fertile mantle [17]. The lateral spreading undergone by this buoyant root is thus expected to behave as a viscous gravity current.

To model this process, the analytical approach of Huppert [34] was used. This model uses a lubrication theory approximation, and describes the shape and rate of propagation of the gravity current. The model assumes a rigid surface; elastic or viscous bending in response to subsurface and surface loads is not included.

According to the model, the height (or thickness) of the swell root, $h(r,t)$, for a given point in the body of the current is:

$$h(r,t) = \xi_N^{2/3}(3Q\nu/g')^{1/4}\, t^{(\alpha-1)/4}\, \psi(z) \tag{3.1}$$

and the radial extent of the current, $r_N(t)$, is

$$r_N(t) = \xi_N(1/3g'Q^3/\nu)^{1/8}\, t^{(3\alpha+1)/8} \tag{3.2}$$

where r is radius, t is time, Q the volume flux, ν the kinematic viscosity, g' the reduced gravity, α the flux rate constant, and ξ_N the value of ξ at $r = r_N(t)$, with

$$\xi = (1/3g'Q^3/\nu)^{-1/8}\, rt^{-(3\alpha+1)/8} \tag{3.3}$$

The starting condition to determine the numerical solution is given by $\psi(z)$ when $z = 1$ $(z = \xi/\xi_N)$

$$\psi(z) = \left[\frac{3}{8}(3\alpha+1)\right]^{1/3}(1-z)^{1/3}\left[1 - \frac{1}{3(3\alpha+1)}(1-z) + O(1-z)^2\right] \tag{3.4}$$

The values chosen for the free model parameters are given in Table 3.1. Using these value-ranges several combinations of parameters were tested in order to find the ones that yielded a swell root shape and dimensions similar to the Cape Verde's swell root (with a radius of 600–800 km, and thickness of 60–80 km). The results are shown in Fig. 3.4.

The corresponding topography for each selected parameter combination is calculated using Lodge's [6] variation of the net swell elevation equation of Morgan et al. [2]:

$$e = h\frac{\Delta\rho}{\Delta\rho_{mrw}} \tag{3.5}$$

where e is the swell topography (m), h is the swell root thickness (m) yielded by the Huppert model, $\Delta\rho$ is the density difference between the swell root and mantle (kg/m^3), and $\Delta\rho_{mrw} = \rho_m - \Delta\rho - \rho_w$, where ρ_m is the mantle density (kg/m^3) and ρ_w is the seawater density (kg/m^3).

Table 3.1 Summary of parameters used in the lubrication model

Symbol	Parameter	Value
ρ_{mantle}	Density of the mantle (fertile)	3,300 kg/m^3
$\Delta\rho$	Density difference between the fertile mantle and the depleted root	$90 \leq \Delta\rho \leq 720$ kg/m^{3a}
η	Depleted swell root viscosity	$1 \times 10^{19} \leq \eta \leq 1 \times 10^{21}$ Pa S^b
α	Flux rate constant	$0 \leq \alpha \leq 1^c$
t	Age of the archipelago	26 Ma^d
Q	Flux	$7.5 \times 10^7 \text{km}^3/\text{Ma}^{\alpha e}$

[a] 90 kg/m^3 is the approximate density difference calculated for a ~65 km thick swell root and the geoid anomaly and 720 kg/m^3 is the density difference inferred by the receiver function analysis [6]

[b] Viscosities are kinematic viscosities (v)

[c] $\alpha = 0$ is equivalent to a single release of a constant volume of swell root, and $\alpha = 1$ is equivalent to a constant flux of material

[d] 25.6 Ma is the oldest numerical age associated with the island volcanism [21], and ~25 Ma is also the age assigned by Faugeres et al. [30] for the uplift of the rise

[e] Calculated using the approximate volume of the swell root/ age of the archipelago, elevated to α; for $\alpha = 0$, Q is $\sim 7.5 \times 10^7 \text{km}^3$; for $\alpha = 1$, Q is $\sim 2.9 \times 10^6 \text{km}^3/\text{Ma}$

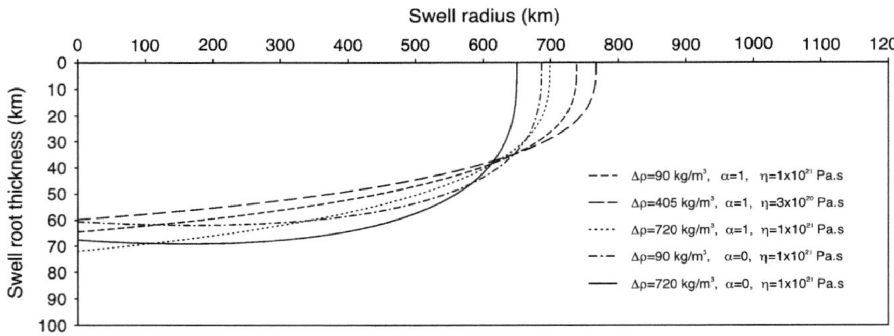

Fig. 3.4 Possible combinations that produce swell root dimensions similar to the Cape Verde swell root (with a radius of 600–800 km and a thickness between 60 and 80 km). $\rho_m = 3{,}300$ kg/m^3, $t = 26$ Ma

The results show that the best fitting parameter combinations are for $\Delta\rho$ near 90 kg/m^3 (see Fig. 3.5). Combinations involving $\Delta\rho$ significantly larger than 90 kg/m^3 yield unrealistic topographies. Furthermore, the results show also that even combinations with a $\Delta\rho = 90\,\text{kg/m}^3$ yield a topography that is slightly overestimated when compared to the observed 1.9–2.3 km height of the swell, suggesting that slightly lower density differences should be used.

The suggested solutions involve a flux rate constant (α) that varies from 0 to 1; $\alpha = 0$ implies instantaneous volumetric release while $\alpha = 1$ corresponds to constant flux conditions. We expect $\alpha \approx 1$ to be more realistic since a single release of material at 26 Ma ago (a situation described by a value of $\alpha = 0$) is not consistent

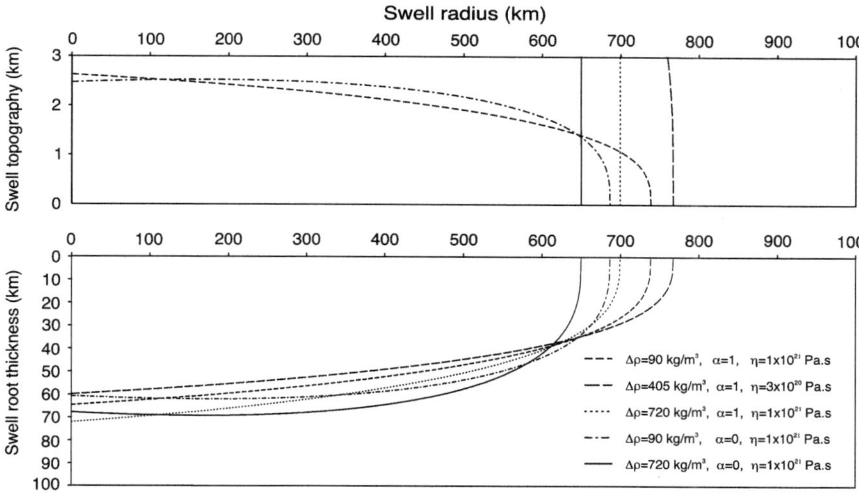

Fig. 3.5 Swell topography for each of the selected combination of parameters. $\rho_m = 3,300\,\text{kg/m}^3$, $\rho_w = 1,030\,\text{kg/m}^3$, $t = 26$ Ma

with a volcanic history extending from that time to the present. Furthermore, an $\alpha \approx 1$ (between 0.7 and 1) would correspond to bursts of discrete releases of material through geological time, which approximates plume behaviour if we consider that its surface expression (island volcanism) is intermittent and the islands exhibit several evolutionary phases. Hence, α within 0.7 and 1 yields characteristics closest to the Cape Verde swell. Morgan et al. [2] estimated an α ranging 0.73–0.77 for the Cape Verde swell.

An additional constraint is the geoid anomaly (ΔN) and the geoid to topography ratio ($\Delta N/H$) associated with the possible numerical solutions. Only solutions that yield $8 \leq \Delta N \leq 12$ m and $3.4 \leq \Delta N/H \leq 6.0$ m/km are in agreement with the observational constraints. In the following calculations ΔN is estimated using a reference column composed of:

$$\underbrace{5.4\,\text{km} @ 1,030\,\text{kg/m}^3}_{\text{sea}} + \underbrace{7\,\text{km} @ 2,800\,\text{kg/m}^3}_{\text{crust}} + \underbrace{87.6\,\text{km} @ 3,300\,\text{kg/m}^3)}_{\text{mantle}} \quad (3.6)$$

The effects on the geoid of the volcanic edifices and the flexural compensation associated with the loading of the edifices were not considered.

After introducing these additional constraints in the model, the range of parameters that yield results in agreement with the observations is narrower. These include $70 \leq \Delta\rho \leq 80\,\text{kg/m}^3$, $0.70 \leq \alpha \leq 1$, and $\eta \approx 1.2 \times 10^{20}$ Pa s. Assuming a best solution with a swell root thickness (h) $\approx$ 65 km, a swell topography (e) $\approx$ 2.0 km (a value suggested by the bathymetric profiles of Fig. 3.6) and a geoid anomaly (ΔN) $\approx$ 8 m, the corresponding parameter combination includes $\Delta\rho \approx 70\,\text{kg/m}^3$, $\alpha \approx 0.70$ and $\eta \approx 1.2 \times 10^{20}$ Pa s (see Fig. 3.7). These values produce a numerical solution in good agreement with the observational constraints.

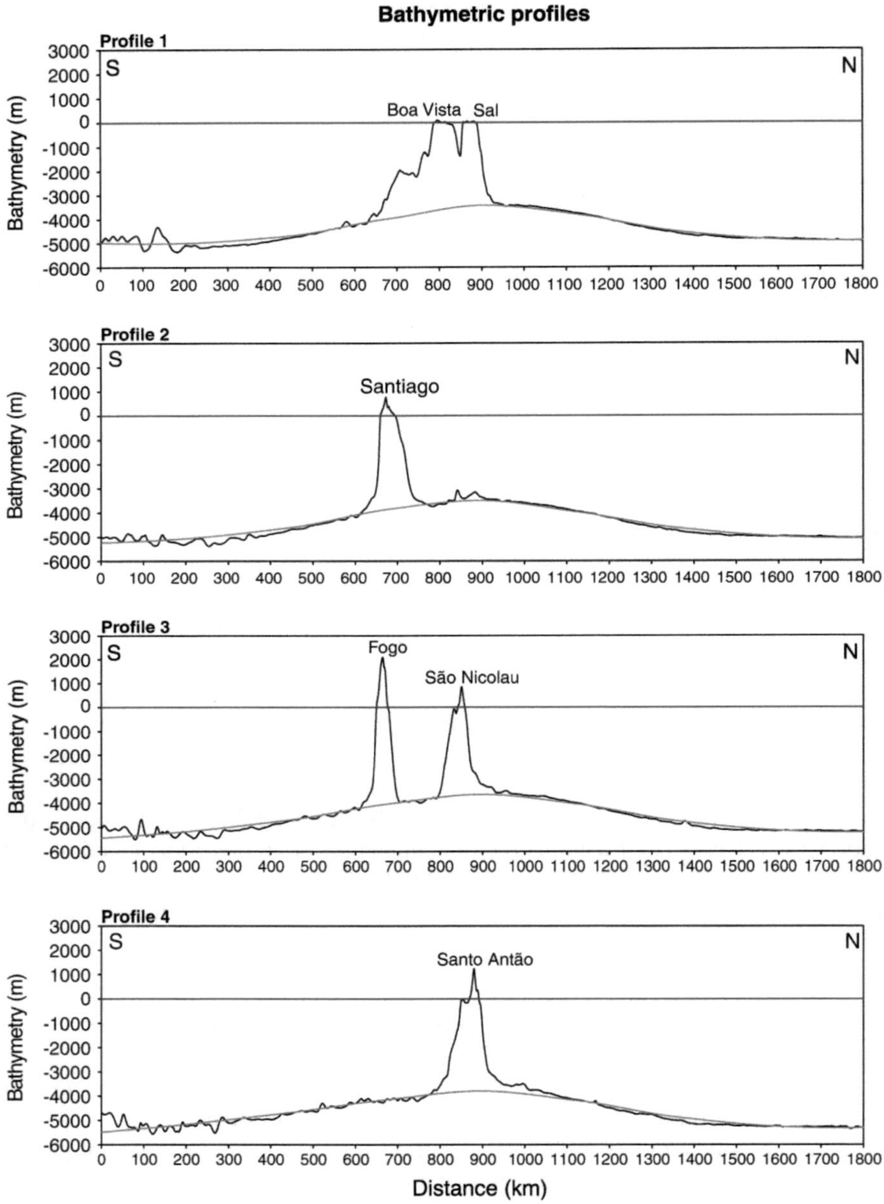

Fig. 3.6 Bathymetric profiles across the Cape Verde Rise, along profiles indicated in Fig. 3.1. Smoothed topography is shown in gray and was calculated using a local polynomial regression fitting function (degree = 2). Elevation data from Amante and Eakins [26]

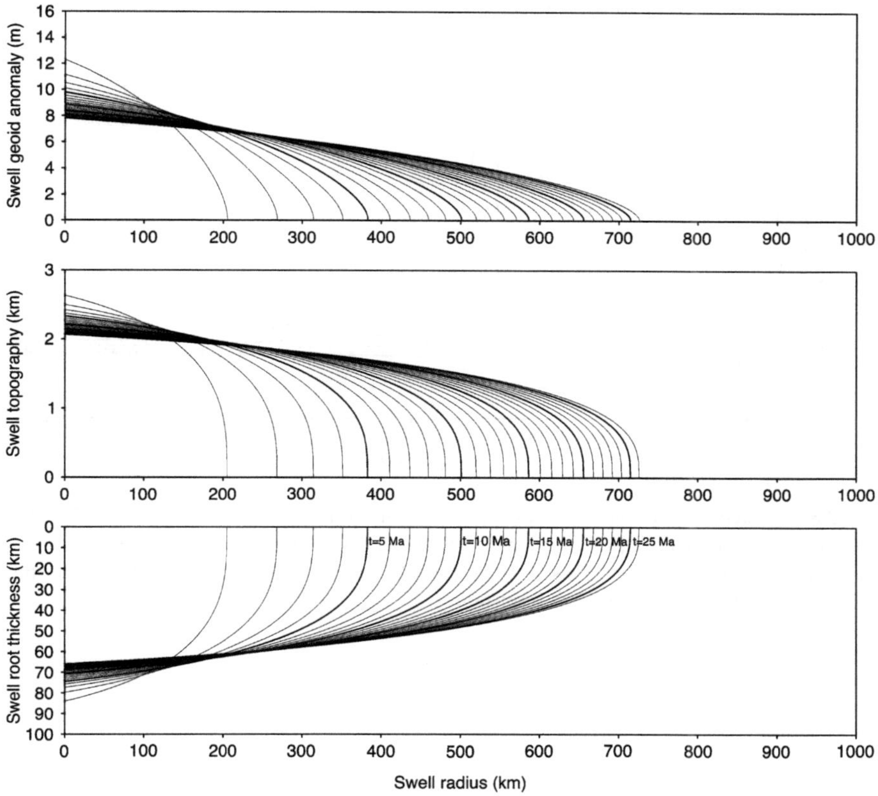

Fig. 3.7 Evolution of the swell root, swell topography and the geoid anomaly over the 26 Ma period using increments of 1 Ma. Bold lines correspond to t = 5, 10, 15, 20 and 25 Ma. Combination of parameters corresponds to: $\rho_{\text{mantle}} = 3,300\,\text{kg/m}^3$, $\Delta\rho = 70\,\text{kg/m}^3$, $\eta = 1.2 \times 10^{20}$ Pa s, $\alpha = 0.70$, $t = 26$ Ma

The evolution of the swell through the 26 Ma period was also computed. Figure 3.7 shows the evolution of the swell root, swell topography and geoid anomaly assuming the axisymmetric spread of a buoyant root below a rigid plate and using Huppert's equations as an approximation to root fluid dynamics. The parameters match the best solution described before. Within the model's approximations, several implications may be inferred: (1) the swell's full height is reached in the initial stages of swell development and is followed by a slight decay tending to stability over time; (2) after the initial uplift the swell's height changes only by a small amount as time progresses while the swell broadening decelerates; (3) at a given point, the seafloor experiences a very rapid uplift followed by a decrease in the uplift rate until a relative vertical stabilisation is slowly achieved with almost no subsequent subsidence or very small subsidence (see Fig. 3.8); (4) the greater the distance between a given point and the swell centre, the later the peak in the uplift rate. Figure 3.8 shows the theoretical instantaneous sea-floor

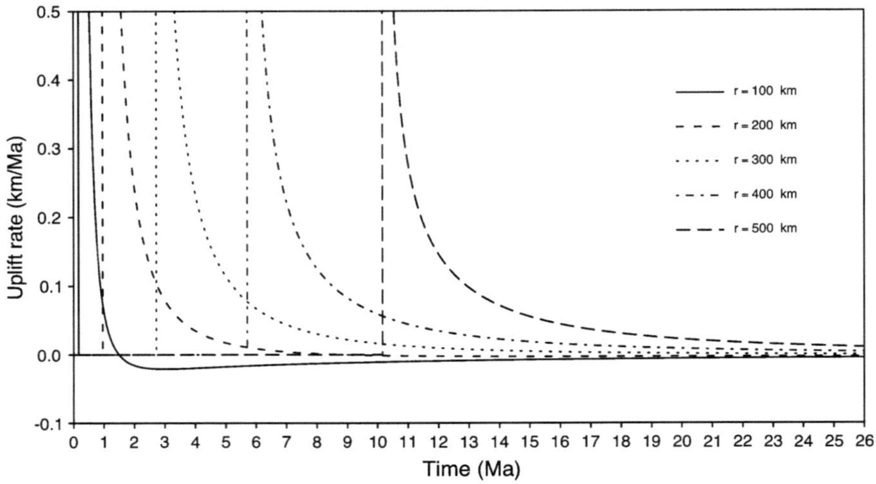

Fig. 3.8 Theoretical instantaneous uplift rates for a given distance from the centre of the swell, calculated using the Huppert [33] equations for swell root development (with $\rho_{\text{mantle}} = 3{,}300\,\text{kg/m}^3$, $\Delta\rho = 76\,\text{kg/m}^3$, $\eta = 1.2 \times 10^{20}$ Pa s, $\alpha = 0.75$) and the resulting topography. The origin of the time scale corresponds to the start of swell formation

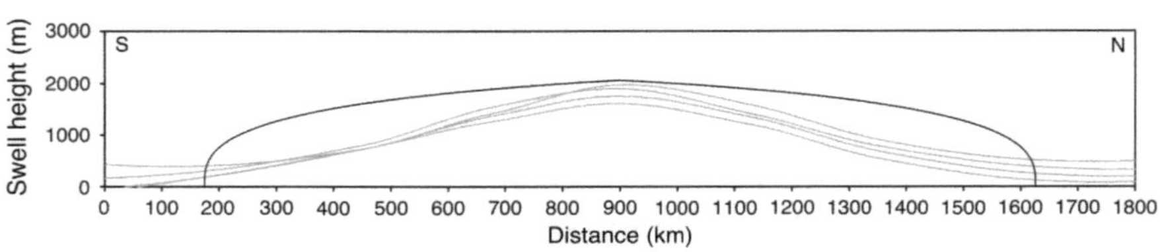

Fig. 3.9 Comparison between the topography of the Cape Verde Swell (smoothed bathymetric profiles in *gray*) and the synthetic topography according to a lubrication theory model (in *black*). Lubrication theory parameters correspond to: $\rho_{\text{mantle}} = 3{,}300\,\text{kg/m}^3$; $\Delta\rho = 70\,\text{kg/m}^3$; $\eta = 1.2 \times 10^{20}$ Pa s; $\alpha = 0.70$; $t = 26$ Ma. Note the misfit between the synthetic topography and the real topography, on the flanks of the swell

uplift rates that occur in the swell region. These show well the vertical stability to which the central area of the swell tends.

The use of a lubrication-theory model coupled with isostatic uplift, however, is not free of problems. Apart from the uncertainties in the definition of the swell dimensions, densities, viscosities and age, we must consider also that the Huppert model is an approximation that deviates from experimental values near the origin, making it problematic to model the very initial stages of evolution of the swell (<1 Ma). Additionally, the model seems to be a poor match for the swell morphology, despite providing a reasonably good fit to the swell's general dimensions, the swell root thickness and the height of the geoid anomaly. The misfit is particularly evident at the propagation front (see Fig. 3.9) where the synthetic topography presents a convex shape while the real swell is characterised by a concave shape with a smooth transition to the background topography. This misfit is mostly due to the fact that the flexural rigidity of the plate is not taken into account in purely isostatic models. Moreover, with respect to the swell evolution simulations, the rigidity of the plate would also have the effect of slowing down the uplift rates. Thus, in order to produce a synthetic topography that fits well the observed swell morphology, one must use a flexural approach.

3.4 Application of a Flexural Model

Lithospheric flexural models treat the lithosphere as a uniform, elastic, thin and spherical shell enclosing a viscous fluid mantle [35]. The rigidity of plates provides a reasonable approach since one of the postulates of plate tectonics is that plates do not deform significantly on geological time scales [36]. A rigid, elastic plate responds to a concentrated surface load over a broader area than the load [37], creating an inner deflected area. Since the swell formation processes seem to be governed by buoyancy forces, this translates into an upward load from below, an approach consistent with the shape of the swell admittance at longer wavelengths [16]. Assuming that the time scale of the loading is on the order of several millions of years, the asthenosphere can be considered a fluid and the load fully compensated [33, 37]. Likewise, the linearity of the elastic equations holds because the displacement is small [16]. Thus, by using a flexural model, we gain insight into the shape of the root of the Cape Verde swell, and possibly others.

In this study a formulation for axisymmetrical loading by Brotchie and Silvester [35] was used. The load is simulated by a stack of disks of decreasing radii, where the contribution of each disk to the deflection of the plate is defined by

$$W(r) = \frac{(\rho_{\text{edifice}} - \rho_{\text{water}})h}{(\rho_{\text{asthenosphere}} - \rho_{\text{moat}})}[A\,Ker'(A)\,Ber(r/l) - A\,Kei'(A)\,Bei(r/l) + 1] \tag{3.7}$$

when inside the loaded area, and

$$W(r) = \frac{(\rho_{\text{edifice}} - \rho_{\text{water}})h}{(\rho_{\text{asthenosphere}} - \rho_{\text{moat}})}[A\,Ber'(A)\,Ker(r/l) - A\,Bei'(A)\,Kei(r/l)] \tag{3.8}$$

when outside the loaded area. $W(r)$ is the deflection of the plate for a given distance from the centre of the load (r), ρ refers to densities, l is the radius of effective stiffness, h is the height of each disk, A = radius of the disk/l, and Ber, Ker, Bei, Kei are zero-order Bessel–Kelvin functions (primed functions correspond to first derivatives).

In order to simulate the loading of a buoyant root from below (see Fig. 3.10), the following arrangements were made to the first term of the equations:

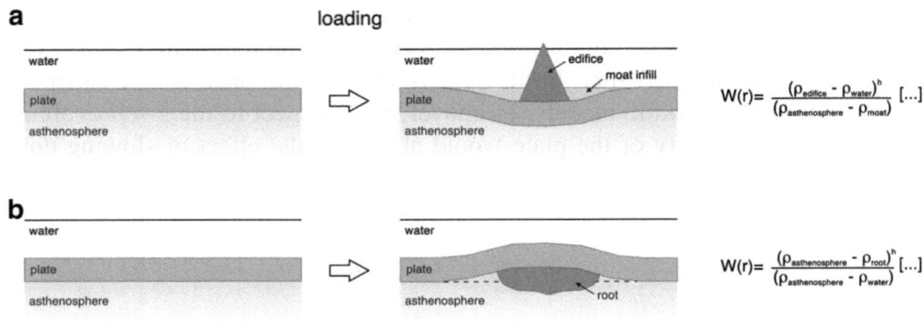

Fig. 3.10 Comparison between the **a** surface loading and **b** subsurface loading models

Table 3.2 Summary of parameters and value ranges used in the flexural model

Symbol	Parameter	Value
D	Flexural rigidity	1.11×10^{24} N m
E	Young's modulus	100 GPa
T_e	Effective elastic thickness	30 km
l	Radius of effective stiffness	7.65×10^4 m
R	Earth's radius	6,371 km
g	Gravitational acceleration	9.81 m/s^2
v	Poisson's ratio	0.25
$\rho_{\text{asthenosphere}}$	Mantle density	3,300 kg/m^3
ρ_{water}	Seawater density	1,030 kg/m^3
ρ_{root}	Swell root density	3,270–3,195 kg/m^3

$$W(r) = \frac{(\rho_{\text{asthenosphere}} - \rho_{\text{root}})h}{(\rho_{\text{asthenosphere}} - \rho_{\text{water}})}[\ldots] \tag{3.9}$$

For the estimation of l, we use

$$l = \left[\frac{D}{\frac{ET_e}{R^2} + \rho_{\text{asthenosphere}}\, g}\right]^v \tag{3.10}$$

where

$$D = \frac{ET_e^{\,3}}{12\,(1 - v^2)} \tag{3.11}$$

A value of 30 km (Table 3.2) for the effective elastic thickness was chosen because this seems to be a robust regional estimation of T_e in the vicinity of the Cape Verde islands [9, 28]. The value of 100 GPa for the Young's modulus was chosen because is an approximate value for old, stiff lithosphere [28]. For the buoyancy parameters the well accepted values of $\rho_{\text{asthenosphere}} = 3{,}300$ kg/m^3 and $\rho_{\text{water}} = 1{,}030$ kg/m^3 were used. For the root density (ρ_{root}) we have tested values ranging from 3,270 to 3,195 kg/m^3, which corresponds to a range of $\Delta\rho$ between 30 and 105 kg/m^3 (from Pim et al. [9]). Additional parameters are also in Table 3.2.

With this set of parameters the swell root was modelled as a stack of disks with decreasing radii as depth increases. The thickness and the position of the centre of each disk was adjusted in order to produce an upward inflection capable of matching the observed smoothed swell bathymetry, thus reflecting the irregularities of the swell as much as the geometry allows. Forward modelling by trial and error was performed in order to find a solution that fits the Cape Verde swell morphological constraints, a compensation depth of $\sim$80 km, and a root density within 3,270–3,195 kg/m^3. The flexure associated with island surface loading was not taken into account.

A subsurface load of $\sim$700 km by $\sim$55 km, for a root density of 3,215 kg/m^3 ($\Delta\rho = 85$ kg/m^3), produces the best fit for the observed swell and is in agreement with the compensation depth. Changes up to 5 km in the swell root thickness or 5 kg/m^3 in $\Delta\rho$ do not alter significantly the solution and thus are still considered acceptable. Likewise, the model is insensitive to <10% changes in the disks' radii. Figure 3.11 shows the synthetic topography produced by the flexural model in comparison with the observed bathymetry. This figure also shows the shape of the approximate root that produces the synthetic profiles. Figure 3.12 shows the position and radius of the stack of disks used to model the swell root shape. The computed swell root shape mirrors the swell shape but at a different scale. Its geometry is different from the root geometry produced by the lubrication-theory model, perhaps suggesting that the Huppert [33] equations in their pure form may not be adequate to describe swell root development.

3.5 Discussion and Conclusions

The use of an analytical model that combines the evolution of a radially-symmetric buoyant current under a rigid plate, together with a purely isostatic model, is a possible approach to swell development on a stationary environment. Forward modelling shows that solutions that include $70 \leq \Delta\rho \leq 80\,\mathrm{kg/m^3}$, $0.70 \leq \alpha \leq 1$, and $\eta \approx 1.2 \times 10^{20}\,\mathrm{Pa\,s}$ are compatible with the Cape Verde swell's observational constraints. These solutions fit well a swell root thickness of 60–80 km, a swell topography within 1.9–2.3 km and a geoid anomaly of $\sim$8 m. Likewise, the density difference between the spreading root and the surrounding mantle ($\Delta\rho$) is also within the observed value range of 3,270–3,195 kg/m^3. The fluid flow equations [34] used to describe the swell root shape also allow the modelling of the swell development over the considered 26 Ma interval. The model suggests that the characteristic signature of a radially-symmetric buoyant spreading root below a rigid surface is a very rapid initial uplift and a subsequent slower rise to a quasi-stationary position with almost no subsequent subsidence. If we consider that the

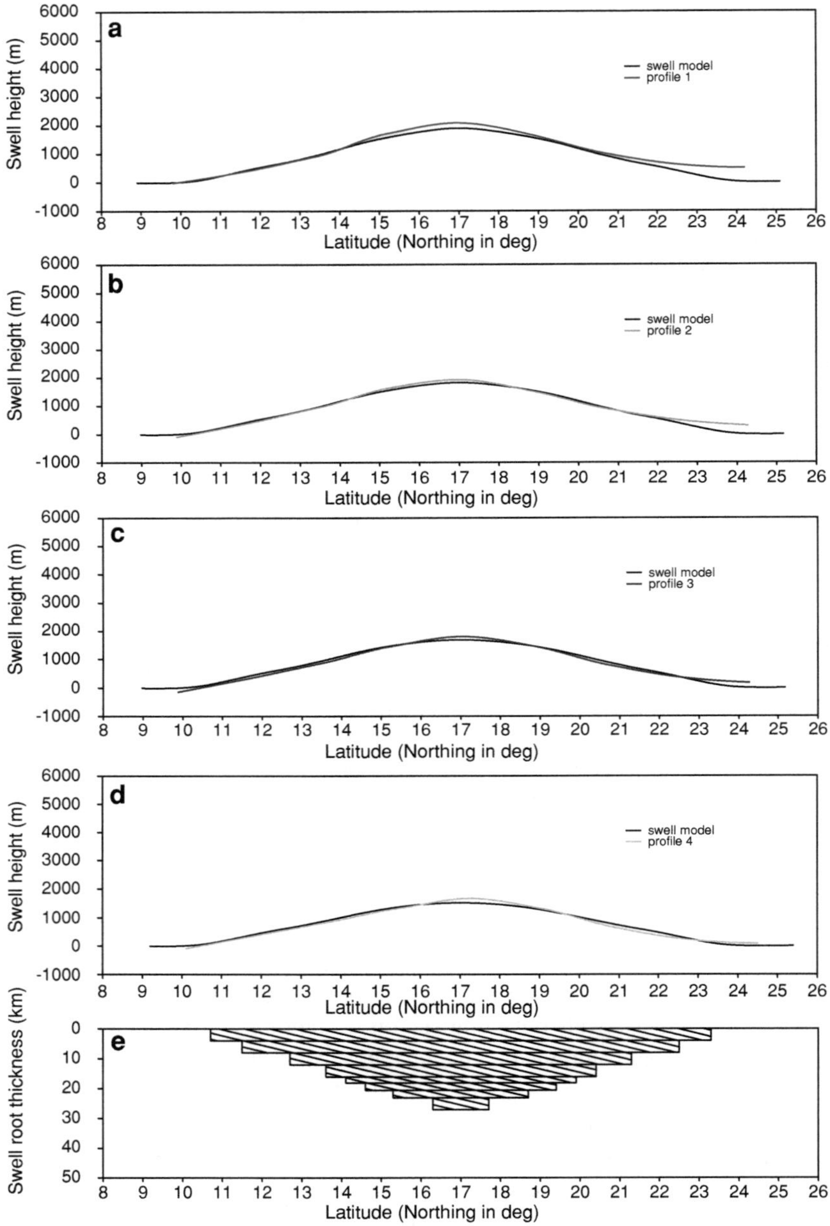

Fig. 3.11 **a–d** Shape of the synthetic topography produced by a subsurface flexural model in comparison with the smoothed observed bathymetry along profiles 1–4. Plot **e** corresponds to the shape of the synthetic swell root at profile 1. Parameters correspond to $T_e = 30$ km, $E = 100$ GPa, $\rho_{\mathrm{root}} = 3{,}215\,\mathrm{kg/m^3}$

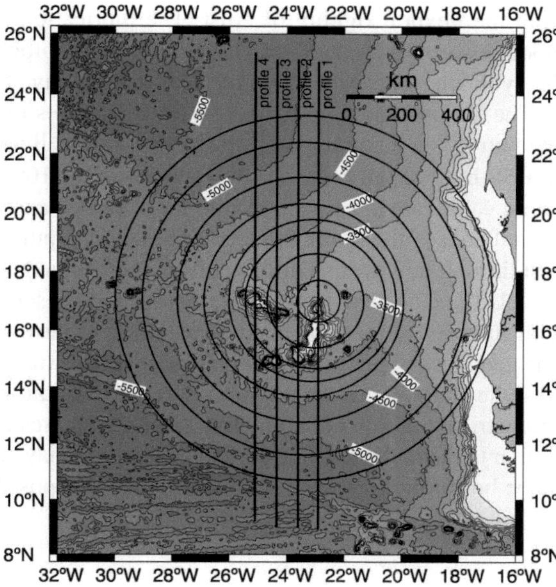

Fig. 3.12 Disk location and radius used in the computation of the synthetic swell root. Note the general agreement between the root geometry and the Cape Verde swell. Profiles 1–4 also shown. Elevation data from Amante and Eakins [26]

islands are normally located within the central area of the swell, this model implies that only the older islands, built in the initial stages of swell development, will hold a record of the initial uplift in their volcanostratigraphies. Islands built within 300 km of the swell centre and within the first 10 Ma of swell development will no longer experience swell-driven uplift if flux conditions are constant (see Fig. 3.8). Conversely, assuming an old lithosphere in thermal equilibrium, islands built after the initial 10 Ma of swell development rest on vertically stable seafloor and only mechanisms unrelated with swell development may account for the hypothetical vertical movements affecting those islands. Another implication of this model concerns the late uplift affecting the seafloor and seamounts situated away from the swell central area. The lateral spreading of a viscous root implies that the seafloor situated on the periphery of the swell experienced uplift significantly later than seafloor located closer to the centre of the swell; this behaviour, if real, must be reflected in seafloor stratigraphy and is potentially trackable. Effectively, the stratigraphic record of IODP site 12 (DSDP 2–12), located $\sim$450 km NNW of the swell central area, may attest to this late uplift since it seems to reflect an abrupt relative lowering of the carbonate compensation depth sometime around the Miocene/Pliocene transition [38, 39]. The timing of this possible uplift is significantly younger than the Late Oligocene/ Early Miocene uplift that occurred in the central area of swell and inferred through the stratigraphy of IODP sites 368 (DSDP 44–368) and 659 (ODP 108–659) [27, 30]. The uplift recorded in the sediments of site 12, however, seems to have occurred at a later time than the time inferred from Fig. 3.7 (at a distance $\sim$450 km from the swell centre), perhaps reflecting the fact that this numerical model does not include the effects arising from the flexural rigidity of crust. The lubrication theory, with a purely isostatic model also offers a poor match to swell morphology, as the comparison between the synthetic topography and the real topography shows.

The flexural model provides better constraints on the swell's hypothetical root geometry. The flexural model shows that the swell root that best fits the topography is a 55 by 700 km root, for $T_e = 30$ km, $E = 100$ GPa, and a $\Delta\rho = 85\,\text{kg/ m}^3$. These values are in agreement with the observed anomalously lower densities in the central area of the swell and a compensation depth ~ 80 km. The geometry of the best fit root, however, is different from the geometry described by the Huppert [34] model. The lack of an appropriate algorithm to describe the evolution of a swell root with such geometry prevents the modelling of the swell over time, for the moment. The geometry of the swell root yielded by the flexural model may be the result of the episodic character of the melting source, a difficult factor to model accurately. The geometric evolution of cumulative discrete releases of material, with an episodicity similar to the main volcanic periods of activity, may be the origin of such a swell root shape but further research is needed to test this possibility.

References

1. Detrick, R., & Crough, T. (1978). Island subsidence, hot spots, and lithospheric thinning. *Journal of Geophysical Research (Solid Earth), 83*(B3), 1236–1244.
2. Morgan, J., Morgan, W., & Price, E. (1995). Hotspot melting generates both hotspot volcanism and a hotspot swell. *Journal of Geophysical Research (Solid Earth), 100*(B5), 8045–8062.
3. Parsons, B., & Sclater, J. (1977). An analysis of the variation of ocean floor bathymetry and heat flow with age. *Journal of Geophysical Research (Solid Earth), 82*(5), 803–827.
4. Stein, C., & Stein, S. (1992). A model for the global variation in oceanic depth and heat flow with lithospheric age. *Nature, 359*(6391), 123–129.
5. Monnereau, M., & Cazenave, A. (1990). Depth and geoid anomalies over oceanic hotspot swells: A global survey. *Journal of Geophysical Research (Solid Earth), 95*(B10), 15–429.
6. Lodge, A. (2006). Seismic constraints on swell formation beneath the Cape Verde Islands (Ph.D. Thesis, University of Bristol).
7. Sleep, N. (1990). Hotspots and mantle plumes: Some phenomenology. *Journal of Geophysical Research (Solid Earth), 95*(B5), 6715–6736.
8. Lodge, A., & Helffrich, G. (2006). Depleted swell root beneath the Cape Verde Islands. *Geology, 34*(6), 449–452.
9. Pim, J., Peirce, C., Watts, A., Grevemeyer, I., & Krabbenhoeft, A. (2008). Crustal structure and origin of the Cape Verde Rise. *Earth and Planetary Science Letters, 272*(1–2), 422–428.
10. Ribe, N. (2004). Through thick and thin. *Nature, 427*(6974), 793–795.
11. Crough, S. (1978). Thermal origin of mid-plate hot-spot swells. *Geophysical Journal International, 55*(2), 451–469.
12. Ribe, N., & Christensen, U. (1999). The dynamical origin of Hawaiian volcanism. *Earth and Planetary Science Letters, 171*(4), 517–531.
13. Davies, G. (1988). Ocean bathymetry and mantle convection 1. Large-scale flow and hotspots. *Journal of Geophysical Research (Solid Earth), 93*(B9), 10467–10480.
14. Olson, P. (1990). Hot spots, swells and mantle plumes. In M. Ryan (Ed.), *Magma transport and storage* (chap. 3, pp. 33–51). Chichester, UK: Wiley.

15. Li, X., Kind, R., Priestley, K., Sobolev, S., Tilmann, F., Yuan, X., & Weber, M. (2000). Mapping the Hawaiian plume conduit with converted seismic waves. *Nature, 405*(6789), 938–941.
16. McNutt, M. (1988). Thermal and mechanical properties of the Cape Verde Rise. *Journal of Geophysical Research (Solid Earth), 93*(B4), 2784–2794.
17. Ribe, N., & Christensen, U. (1994). Three-dimensional modeling of plume-lithosphere interaction. *Journal of Geophysical Research (Solid Earth), 99*(B1), 669–682.
18. Harris, R., & McNutt, M. (2007). Heat flow on hot spot swells: Evidence for fluid flow. *Journal of Geophysical Research (Solid Earth), 112*(B3), B03407.
19. Sandwell, D., & Mackenzie, K. (1989). Geoid height versus topography for oceanic plateaus and swells. *Journal of Geophysical Research, 94*(B6), 7403–7418.
20. Li, X., Kind, R., Yuan, X., Wölbern, I., & Hanka, W. (2004). Rejuvenation of the lithosphere by the Hawaiian plume. *Nature, 427*(6977), 827–829.
21. Torres, P., Silva, L., Serralheiro, A., Tassinari, C., & Munhá, J. (2002). Enquadramento geocronológico pelo método K/Ar das principais sequências vulcano-estratigráficas da Ilha do Sal—Cabo Verde, Garcia de Orta. *Serviços Geológicos, 18*(1–2), 9–13.
22. Burke, K., & Wilson, J. (1972). Is the African Plate stationary? *Nature, 239*(5372), 387–390.
23. Morgan, W. J. (1983). Hotspot tracks and the early rifting of the Atlantic. *Tectonophysics, 94*(1–4), 123–139.
24. Gripp, A., & Gordon, R. (2002). Young tracks of hotspots and current plate velocities. *Geophysical Journal International, 150*(2), 321–361.
25. Crough, S. (1982). Geoid height anomalies over the Cape Verde Rise. *Marine Geophysical Researches, 5*(3), 263–271.
26. Amante, C., & Eakins, B. W. (2009). ETOPO1 1 arc-minute global relief model: Procedures, data sources and analysis.*NOAA Technical Memorandum NESDIS NGDC-24,*19.
27. Lancelot, Y., Seibold, E., Cepek, P., Dean, W., Eremeev, V., Gardner, J., et al. (1978). Site 368: Cape Verde Rise. In Y. Lancelot, E. Seibold, P. Cepek, W. Dean, V. Eremeev, J. Gardner, et al. (Eds.), *Initial reports of deep sea drilling project* (vol. 41, pp. 233–326). Washington: U.S. Government Printing Office.
28. Ali, M. Y., Watts, A. B., & Hill, I. (2003). A seismic reflection profile study of lithospheric flexure in the vicinity of the Cape Verde Islands. *Journal of Geophysical Research (Solid Earth),108*(B5), 2239–2263.
29. Lancelot, Y. (1978). The evolution of the Central Northeastern Atlantic—Summary of results of DSDP Leg 41. In Y. Lancelot, E. Seibold, P. Cepek, W. Dean, V. Eremeev, J. Gardner, et al. (Eds.), *Initial reports of deep sea drilling project* (vol. 51, pp. 1215–1245). Washington: U.S. Government Printing Office.
30. Faugeres, J. C., Legigan, P., Maillet, N., & Latouche, C. (1989). Pelagic, turbiditic, and contouritic sequential deposits on the Cape Verde plateau (leg 108, site 659, Northwest Africa): sediment record during Neogene time. In *Proceedings of the Ocean Drilling Program, Scientific Results* (vol. 108, pp. 311–328).
31. Courtney, R., & White, R. (1986). Anomalous heat flow and geoid across the Cape Verde Rise: evidence for dynamic support from a thermal plume in the mantle. *Geophysical Journal International,87*(3), 815–867.
32. Monnereau, M., & Cazenave, A. (1988). Variation of the apparent compensation depth of hotspot swells with age of plate. *Earth and Planetary Science Letters,91*(1–2), 179–197.
33. Oxburgh, E., & Parmentier, E. (1977). Compositional and density stratification in oceanic lithosphere-causes and consequences. *Journal of the Geological Society,133*(4), 343–355.
34. Huppert, H. (1982). The propagation of two-dimensional and axisymmetric viscous gravity currents over a rigid horizontal surface. *Journal of Fluid Mechanics,121*(1), 43–58.
35. Brotchie, J. F., & Silvester, R. (1969). On crustal flexure. *Journal of Geophysical Research (Solid Earth), 74*(22), 5240–5252.
36. Turcotte, D. L., & Schubert, G. (2002). *Geodynamics* (2nd ed.). Cambridge, UK: Cambridge University Press.

37. McNutt, M., & Menard, H. W. (1978). Lithospheric flexure and uplifted Atolls. *Journal of Geophysical Research (Solid Earth), 83*(B3), 1206–1212.
38. Peterson, M., Edgar, N., von der Borch, C., Cita, M., Gartner, S., Goll, R., & Nigrini, C. (1970). Site 12. In M. Peterson, N. Edgar, M. Cita, S. Gartner, R. Goll, C. Nigrini, & C. von der Borch (Eds.), *Initial reports of deep sea drilling project* (vol. 2, pp. 249–303). Washington: U.S. Government Printing Office.
39. Peterson, M., Edgar, N., von der Borch, C., Rex, R. (1970). Cruise leg summary and discussion. In M. Peterson, N. Edgar, M. Cita, S. Gartner, R. Goll, C. Nigrini, & C. von der Borch (Eds.), *Initial reports of deep sea drilling project* (vol. 2, pp. 413–427). Washington: U.S. Government Printing Office.

Chapter 4
How to Trace Island Freeboard

4.1 Introduction

Changes in island freeboard are an invaluable probe of the processes governing plume/plate interaction and their surface manifestations. However, the study of island freeboard must start from its effects rather than its causes.

The vertical movements affecting island edifices have a direct effect on the evolutionary history of those edifices. During the seamount stage, the height of the edifice relative to the ocean's surface—a parameter that mostly depends on the volcanic activity and the uplift/subsidence of the edifice—has a strong impact on the dominant eruptive style. This happens because the height of the water column above the edifice, hence the hydrostatic pressure, controls the effusive/explosive character of the volcanism. Later, during the emergent or subaerial shield building stages, island freeboard also influences the type of lavas produced and the morphology of the edifice. Furthermore, the vertical trend directly influences the erosive regime affecting the island edifice, a factor that becomes more important towards the end of the volcano's eruptive life. Finally, uplift (or the absence of uplift) will determine whether the edifice remains an island or becomes a guyot.

In order to reconstruct island freeboard, one must find a set of tracers that track a geodetic datum through time. We consider as tracers of uplift and subsidence any geological features that, relatively to a reference level (datum), allow the quantification of the vertical movements affecting the particular geological structure under study (the Cape Verde Islands in our case). There are three main reference data that can be used to trace uplift/subsidence: (1) sea-level, (2) the top of the plate prior to flexure, and (3) the distance to an external reference (e.g. a satellite). Each of these present their own advantages and disadvantages.

First, consider sea-level. Sea-level is normally the reference used for most uplift studies, especially on ocean islands and continental shorelines. This is mainly because sea-level provides a more tangible reference that can be used worldwide, with provisos. Another reason to use sea-level as a reference datum concerns the fact that evidence for past sea-level positions constitute some of the

R. A. S. Ramalho, *Building the Cape Verde Islands*, Springer Theses,
DOI: 10.1007/978-3-642-19103-9_4,

most easily recognisable features—features that are dateable and allow a finer-scale reconstruction of the uplift history. However several problems are associated with this base level. First, the sea-level is not a completely uniform surface as initially thought [1]. Changes in the geoid, salinity, temperature or simply the local gravity field may cause irregularities in the sea-level surface. These irregularities can be up to a few meters and might affect uplift estimates if they require that sort of precision. Second, more importantly, it is known that sea-level has changed over geological time [2–4]. We now have a reasonable record of this change through existing eustatic curves. However these curves are generally discontinuous and strongly depend on the proxies used to estimate the sea-level height. In addition, and for the reasons previously outlined, there is still a large uncertainty in the eustatic curves: 10–50 m or possibly even more [4].

Consider next the top of the plate as a reference datum. To use this as a reference position has the advantages of allowing a better constraint on plate flexure, and avoiding the uncertainties associated with sea-level. However, for one to be able to use the top of the plate as a reference datum, it becomes necessary to accurately define the position of the plate prior to flexure. This is normally achieved through seismic profiles across the wavelength of the flexure to accurately define where the undeformed plate away from the load starts, and thus to determine the origin of the datum. The exercise of finding the position of the undeformed plate to be used as a reference is also dependent of sediment artifacts and the "removal" of the hotspot swell effects. In addition, the seismic imaging is also necessary to constrain the moat's stratigraphy in the search for geometrical evidence for uplift/subsidence. Thus, the reconstruction of an uplift history can only be done using the geometrical relationships within the moat's stratigraphic sequence, and is dependent on the quality of the seismic imaging. Furthermore, precise uplift timings are unavailable unless there is accurate dating of the moat's stratigraphic levels. In a similar fashion, if a complex history of multiple loading occurred, it becomes extremely complex to reconstruct the events that took place to produce the observed final geometry. Notwithstanding these caveats, to constrain the geometry of the plate's flexure and the moat's stratigraphy provides a useful and invaluable tool to investigate the causes of uplift.

Finally, consider the use of an external reference. Small changes in the Earth's surface elevation can nowadays be measured with modern satellite techniques like synthetic aperture radar (SAR) interferometry [5–7]. This technique measures phase differences in aligned SAR images of the same surface area, detecting small changes over time. However this technology can only detect movements between images and cannot be used to look into the geological past. Another technique that uses a reference level is GPS survey. However, like SAR, this technique cannot be used beyond the first survey date.

We chose to use the sea-level as a reference datum because we believe its advantages surpass the eventual disadvantages. The uncertainty in the eustatic curve is the major disadvantage. However, the focus of our study is ocean islands and inevitably their evolution is controlled by sea-level position through its influence on the eruption style and the dominant erosion mechanisms.

Moreover, the stratigraphic record of the islands also retains dateable evidence of past sea-level positions, even if discontinuous. Hence, sea-level and sea-level related uplift tracers, allow long-term vertical movement reconstructions with a direct link to the islands' evolutionary histories.

Since we opted to use global sea-level as a reference datum for our vertical movement reconstructions, the kind of tracers of uplift and subsidence we use are palaeo-markers of sea-level. We consider as palaeo-markers of sea-level any geological formation that recorded the relative position of sea-level during its formation. We will use hereafter the concept of "tracers of uplift and subsidence" interchangeable with "palaeo-markers of sea-level" since our reference datum to trace island freeboard is sea-level. However, we would like to highlight that these two concepts are different in principle. Palaeo-markers of sea-level are just one of the possible tracers that can be used to infer uplift and/or subsidence.

4.2 The Concept of Vertical Displacement

The quantification of island freeboard is based on the simple concept of vertical displacement. In this context, the vertical displacement is defined as the long-term shift of land in a vertical direction, resulting in a change in elevation. The determination of vertical displacement involves the comparison between the present elevation of a particular marker and the elevation of a reference datum during a time correspondent to the age of the marker. If the vertical displacement occurred in the downwards direction, i.e. if the marker ended with a relative elevation lower than the reference datum, we say the marker experienced subsidence. Conversely, if the vertical displacement occurred in the upwards direction, i.e. if the marker ended with a relative elevation higher than the reference datum, we say the marker experienced uplift. If the displacement is null (equal to 0), we say the marker is stationary relative to the reference datum.

Numerically, and since our reference datum is sea-level, the vertical displacement of a particular marker is expressed by the following equation (see also Fig. 4.1):

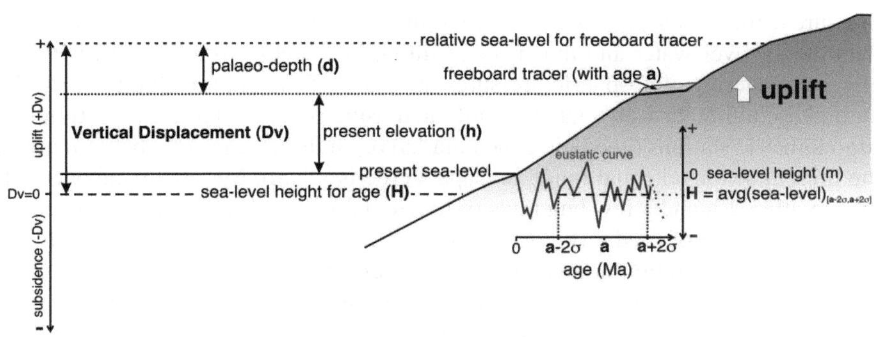

Fig. 4.1 The determination of vertical displacement through palaeo-markers of sea-level

$$D_v = h + d - H \tag{4.1}$$

where D_v is vertical displacement, h is the present elevation, d the inferred palaeo-water depth, and H is the contemporaneous sea-level height extracted from a eustatic curve. Both h and H may be either positive (+) or negative (−) depending on whether, respectively, h refers to a marker's elevation above or below the present sea-level, and H corresponds to a past global sea-level above or below the present level. In contrast, d is always positive (+) since the depth of a marker will always place the sea-level above that marker. Consequently, D_v is positive if the movements affecting the marker resulted in uplift, negative if resulted in subsidence, and zero if the marker is stationary or if the combination of movements affecting the marker resulted in a position identical to the initial one. Since each marker represents only a snapshot in the geological history of the island and does not record the exact vertical movement path undergone by that island, the estimated D_v is the final product of a range of possible solutions that resulted in a particular displacement. Hence, D_v is a resulting vertical displacement. However, a large number of available palaeo-markers will allow a reconstruction that will approximate the long-term vertical motion path affecting the edifice. Likewise, it will also allow the estimation of displacement rates through an appropriate regression model.

In the following sections we will present and discuss how we determine h, H and d, and their associated errors. We will also present a complete outline of the most common sea-level markers found on ocean islands, available for use in freeboard studies.

4.3 Determination of the Present Elevation

The present elevation (h) of the studied palaeo-markers of sea-level was determined with a handheld GPS receiver with digital barometric altimeter (Garmin GPSMAP 60CS), in conjunction with a 1/25,000 topographical map with a 10 m contour interval. According to the manufacturers specifications, the barometric altimeter has a 0.3 m (1 ft) resolution.

Ocean islands are subject to a well-known bimodal and diurnal atmospheric pressure drift. This occurs because during the day the air over dry land heats faster than the air over water and thus rises resulting in a breeze from the sea, with the effect of lowering atmospheric pressures over land. To avoid this diurnal drift, the barometric altimeter was regularly calibrated using points of known elevation like trigonometric stations (geodetic bench marks) or, in the absence of a better marker, the present sea-level (tidal range in Cape Verde is <1 m). Measurements were all made within 2 h and a few hundreds of horizontal meters from a calibration point. We empirically assigned errors in the determination of the elevations as ± 1 m for stations near calibration points and ± 5 m for distant ones.

The elevations of sedimentary features presented in this work are referenced to the base of the deposits, unless stated otherwise. When elevation ranges are presented, the thicknesses of the deposits are taken into account.

4.4 The Eustatic Level as a Reference Datum

The global mean sea-level is termed eustatic sea-level [8], and it is measured relative to some fixed point such as the centre of the Earth [9]. Its definition excludes short-term variations like tidal or wave oscillations [10]. Despite being termed "global", the eustatic level is not, in fact, an entirely uniform and permanent horizontal surface [1]. Regional changes in salinity, temperature and the geoid affect the sea-level surface up to a few meters [1, 10]. The global mean sea-level provides a practical reference datum to track land movements. However, the position of sea-level recorded by any land-based marker (geological or manmade) is the result of local vertical land movements and/or global sea-level changes in the fixed frame [11, 12]. Thus, the term relative sea-level or local sea-level is used in contrast with eustatic sea-level. The first refers to the local position of sea-level without removing the tectonic component, i.e. it is measured between the sea-surface and a local moving datum, while the latter refers to the worldwide absolute sea-water level. The eustatic sea-level thus provides a datum from which upward or downward movements of relative sea-level can be measured [10]. In consequence, the position of the relative sea-level is expressed in meters above (eustatic) sea-level (asl in abbreviation), or in meters below (eustatic) sea-level (bsl in abbreviation).

Sea-level changes occur on a variety of different scales: from small and short-term scales (tides, waves, seasonal tides, etc), to long-term scales that range between 10^3 and 10^8 years and from a few meters to up to 350 m [3, 4, 9, 12]. Consequently, eustatic sea-level has fluctuated throughout geological time [2–4, 11, 12]. These fluctuations result from the combined effect of changes in the volume of water of the oceans or changes in the volume of the ocean basins [4]. The changes in the volume of water of the oceans have two distinct signatures: a high frequency, high amplitude change associated with the growth and decay of continental ice sheets, and a high frequency, low amplitude change associated with a myriad of processes that include thermal expansion/contraction of seawater, variation in lake storage, etc. [4]. The enhanced cycles of expansion and contraction of continental ice sheets during the Quaternary are, for example, responsible for the very rapid and extreme sea-level changes during this period [9]. In contrast, changes in the volume of the ocean basins are normally linked with long-term changes. Some of the mechanisms that can change the volume of ocean basins include: variations in sea-floor spreading ridges and ridge lengths, rates of sedimentation, and rates of emplacement of oceanic plateaus [2, 4]. The fluctuations associated with these mechanisms vary from very high amplitude (when caused by ridge related processes) to low amplitudes, but at always at slow rates (see Table 4.1). Hence, the resulting eustatic level reflects the combined effects of the different mechanisms, and will also be characterised by changes of different orders (scales).

The construction of a global sea-level model—termed a eustatic curve—has been the focus of variety of studies for the past three decades. One of the first

Table 4.1 Timing and amplitudes of geological mechanisms of eustatic change [4]

Geological mechanism	Amplitude	Rate
Growth and decay of continental ice sheets	Up to 200 m	Up to 20 m/ka
Thermal expansion, variation in groundwater and lakes	5–10 m	10 m/ka
Variation in sea-floor spreading rate or ridge length	100–300 m	10 m/Ma
Variation in sedimentation	60 m	10 m/Ma
Emplacement of oceanic plateaus	–	10–60 m/Ma

techniques employed to reconstruct the fluctuations in sea-level through geological time used detailed studies of sequence stratigraphy of offshore sediments [3, 9]. This technique is called sediment backstripping, and is an inverse technique that allows the extraction of sea-level change amplitudes from the stratigraphic record, taking into account diagenetic sediment compaction [4]. Since the stratigraphic record is discontinuous and incomplete, the resulting curves will inevitably "smooth" some of the high frequency changes, and will have periods in which sea-level is arbitrarily interpolated between known points. However, this is still the only practical method to infer sea-level height from the Miocene backwards. Another technique, that gained prominence recently, is based in the glacioeustatic isotopic proxy δ^{18}O, that enables the reconstruction of sea-level through the chemistry of benthic foraminifera [4, 13, 14]. This technique, despite its uncertainties, permits detailed reconstruction of the high frequency changes during the last 6–7 Ma [4, 13, 14]. The study of tropical reefs and atolls also provides reliable geological estimates [4, 15, 16], and has gained importance in the last two decades with the advent of more accurate radiometric dating. However, the information extracted from this record is limited by the dating techniques to the last few hundred thousand years. Also, the information is poor regarding lowstands, and many of the reefal structures are located in places with complex uplift/subsidence histories [4]. This method, nevertheless, may provide a useful reference frame to calibrate other methods.

The choice of one eustatic curve over another is not straightforward. Each eustatic curve will depend on the quality of the portion of the stratigraphic record used in its reconstruction, but will also depend on the tectonic stability of the location where the study was done. Likewise, it will depend on the methods utilised and the quality of the data. Currently, two eustatic curves existing in the literature stand out due to their long range reconstructions (Phanerozoic): the curves by Haq et al. [3] and by Miller et al. [4]. Since the work described here may include ages spanning from the present to 26 Ma (the age of the oldest exposed rock in the Cape Verde Archipelago [17]) one of these curves will have to be used. The curve provided by Haq et al. [3] is completely based on backstripping techniques while the more recent curve by Miller et al. [4] combines data from δ^{18}O eustasy (in the last 7 Ma) and backstripped data throughout the Miocene. We have opted to use the Miller et al. [4] curve for several reasons. First, its more recent date reflects better knowledge and more data associated with eustatic reconstruction. Second, because it uses δ^{18}O inferred sea-level it reflects the high frequency changes that occurred during the the last 7 Ma, allowing more robust age uncertainties. Third, it reconciles two datasets (oxygen isotopic and back-stripped) in one single curve. Another advantage of this curve is the accompanying lookup table with the numerical values used in the curve, facilitating plotting and computation. One of the problems arising from the use of the Miller et al. [4] curve is that the extreme values might be poorly constrained.

The Miller et al. [4] curve is presented in a lookup table with increments of 0.005 Ma over the last 7 Ma, and with increments of 0.1 Ma throughout the Miocene. According to Miller et al. [4], the error associated with the eustatic curve is between ± 10 and ± 50 m on the 10^6 year scale. For our study we decided to neglect these errors because we have little control on their variability over time.

4.5 Palaeo-markers of Sea-level

The bulk of the palaeo-markers of sea-level found on ocean islands can be classified as three main types according to their nature: (1) submarine volcanic morphologies, (2) marine sedimentary formations, and (3) erosional morphologies. We will briefly outline their main characteristics and provide suitable examples from the Cape Verdes and other ocean islands. We will also explain how can we use these features in order to reconstruct island freeboard.

4.5.1 Submarine Volcanic Morphologies

Volcanic eruptions occurring in a subaqueous environment, or subaerial eruptions whose products have reached a water body, produce a variety of distinct morphologies [18–21], namely: pillow-lavas, lobate lavas, hyaloclastites, submarine sheet flows, etc. The major factors controlling the generation of these morphologies are: (1) the composition of the magma (and the magma's volatile content), (2) the eruption rate, (3) the slope angle and morphology, and (4) the hydrostatic pressure of the surrounding environment [21–23]. These morphologies, when present in a volcanic succession, indubitably indicate the subaqueous environment in which they were formed. They may also give more specific clues about the prevailing environmental conditions during emplacement. Hence, subaqueous volcanic morphologies may be used to infer the position of the water level contemporaneous with their formation. However, even on ocean islands, an underwater environment does not necessarily mean a marine environment. The geological context in which the morphologies occur is crucial to decide if they were formed in the sea or, alternatively, in a lake or stream. This is particularly important when dealing with isolated or spatially limited outcrops. However, on ocean islands and especially in arid climates (like the climate of Cape Verde), large outcrops or laterally extensive outcrops of subaqueous lavas normally indicate marine environments. We shall describe the principal submarine volcanic morphologies that are likely to be found on ocean islands, focusing on effusive rather than explosive products since these are predominant in the dominantly basaltic island edifices.

4.5.1.1 Pillow and Lobate Lavas

Pillow-lavas are the most common kind of submarine lavas and are so called because they consist of roughly spherical, bulbous, or tube-like forms that resemble pillows [23]. They occur in a variety of shapes and sizes, ranging from a few tens of centimeters to several meters [23–25]. Their outer surface is normally rough, glassy, with polygonal cooling joints and longitudinal grooves. In cross section they normally present radial and concentric cooling joints. These are all features that result from extrusion, quenching and cooling in an underwater environment.

Pillow-lavas normally form a network of interconnected tubes because the slow effusion rates, and/or extremely low viscosities, and the restriction of flow by rapid growth of chilled crusts makes them advance as small digital lobes [26]. If the underlying bottom is steeply dipping, pillow-lavas assume a more linear tube or ribbon-like morphology, oriented downslope. When close to the feeding vent, pillow-lavas frequently form prismatic accumulations called pillow mounds or "haystacks", or pillow volcanoes when they reach a considerable size. When the volcanism occurs along a fissure, effusive eruptions form linear steep-sided pillow ridges [23]. Megapillows are large inflated pillow-like lavas, up to several meters in diameter, considered by some authors to be the feeding system of several stems of smaller pillow-lavas [21, 23]. Figure 4.2 shows examples of pillow-lavas.

Lobate lavas resemble large flattened pillow-lavas with smoothed or polygonally cracked glassy surfaces and frequently hollow interiors [23]. These morphologies are not as frequent on ocean islands as they are at mid ocean ridges. Their genesis is usually associated with fast flow rates and inflation/deflation phenomena [23].

4.5.1.2 Pillow Breccias, Hyaloclastites and Other Hydroclastic Products

The extrusion and growth of pillow-lavas is subject to nonexplosive processes like quenching fragmentation, spall fragmentation and impact shattering [23]. This is particularly true when the extrusion occurs along steep slopes, causing the rolling over and shattering of individual pillows. As a result, variable amounts of mixed material are produced, comprising broken or whole pieces of pillow-lavas enveloped in sand-size matrix of small glass shards. The resulting rock is called pillow breccia and the smaller glassy fragments called hyaloclastites [23]. The interaction between hyaloclastites and water is frequently subjected to a typical alteration process that leads to the formation of palagonite—a hydrated alteration product with a distinct yellow-orange colour.

Pillow breccias and hyaloclastites normally form poorly sorted deposits lying in prograding gently or steeply dipping foresets, usually intercalated with tube-like downsloping pillow-lavas. The generation of deposits with these characteristics is usually associated with steep bottom morphologies like the outer flanks of seamounts and peripheral submarine slopes of ocean islands. Hence, it is not surprising that large deposits of this type often crop out in the submarine series of uplifted ocean islands (e.g. in São Nicolau and Santiago, in Cape Verde, or in Santa Maria, Azores). Examples of these deposits are shown in Fig. 4.2.

Fig. 4.2 Examples of pillow-lavas and hyaloclastites from Cape Verde and the Azores. **a** Pillow-lavas over coral reef, Santiago. **b** Pillow-lavas from São Nicolau. Note the brown palagonitic crust generated from the alteration of the volcanic glass. **c** Steeply dipping foresets of hyaloclastites and ribbon-like pillow-lavas, Santa Maria. **d** Pillow breccia from São Nicolau. Note the brown palagonitic alteration of the hyaloclastitic matrix. **e** Detail of the surface of a submarine lava, São Nicolau. Note the bulbous glassy texture. **f** Large scale foresets of pillow-lavas (below *dashed line*), dipping seaward and generated by island flank growth, Santa Maria

Underwater explosive eruptions also create hydroclastic materials and submarine pyroclasts. These normally result in a complex mix of vesiculated hydroclastic tephra and hyaloclastites [26, 27]. As a general rule, submarine explosive volcanism only occurs when the the magma's volatile pressure exceeds hydrostatic pressure [21, 28–30]. Since the magma's volatile content is variable, the water depth at which explosive eruptions can occur (called the pressure compensation level or the critical depth for explosive volcanism) is also variable [21, 30]. Despite the fact that evidence for explosive volcanism has been found at greater depths, the critical depth for explosive volcanism is probably between 3,000 and 200 m and frequently above 1 km [21, 27–29]. Hence the generation of this kind of deposits is more frequent when submarine volcanic edifices shoal to reach an intermediate water depth [27]. A good example of deposits of this type can be found in the Figueira de Coxe Formation in São Nicolau, in the vicinity of Ribeira Brava village [31].

In shallow waters (<200 m) the enhanced nucleation of bubbles and generation of massive quantities of steam increase the explosivity index and the height of the eruptive column, resulting in vigorous surtseyan eruptions [21, 22, 27]. The resulting phreatomagmatic deposits are very distinctive but vary in characteristics according to the distance to the vent [22]. The deposits vary from poorly sorted or bimodally sorted, poorly structured deposits in the more proximal area, to well stratified planar-bedded to cross-laminated deposits of fine pyroclasts in more distal areas [21, 22]. These kind of deposits are frequent on ocean islands and are more frequently generated during the emergent phase and during the rejuvenated stage. Phreatomagmatic eruptions may also occur inland, when magma interacts with groundwater or freshwater bodies. However, this is unlikely in arid conditions like those prevailing in Cape Verde. One way to confirm this is to look for marine fossils in the distal deposits, since marine shelly organisms are sometimes entrained and deposited by the phreatomagmatic activity.

4.5.1.3 Submarine Sheet Flows

Submarine sheet flows are also called massive submarine flows or simply sheet flows. They are the result of high flow rates [21, 23, 24, 32, 33]. As their name suggests, they consist of broad sheet-like lava flows that advance rapidly and cover large areas. They are normally unvesiculated and phenocrysts are rare. This type of lava flow is well known from mid ocean ridge environments [34, 35]. Only recently authors started to realise that they are also frequent in seamounts [21, 32, 33]. Now it is thought that up to 20–30% of all submarine lava eruptions form sheet flows [21]. As a result of their "recent recognition", sheet flows from seamounts and ocean islands are still poorly described, and transitional morphologies are reduced to a few small marginal mentions. The detailed study of sheet flows is beyond the scope of this work. However, their main characteristics shall be described because they are frequent features in the submarine series of Santiago and São Nicolau, in Cape Verde, and in Santa Maria, in the Azores.

The observed sheet flows from the Cape Verdes and the Azores exhibit a variety of surface morphologies ranging from smooth, lineated, folded, jumbled, or more frequently with jigsaw cooling joints. In cross section they exhibit a range of distinct morphologies that include (see Fig. 4.3):

- curvicolumnar jointing and columnar jointing characterised by contorted, closely packed columns with glassy undulating or crenulated faces;
- hackly and cubic "kubbaberg" jointing—a form of entablature columnar jointing that resembles box-joints [36–39];

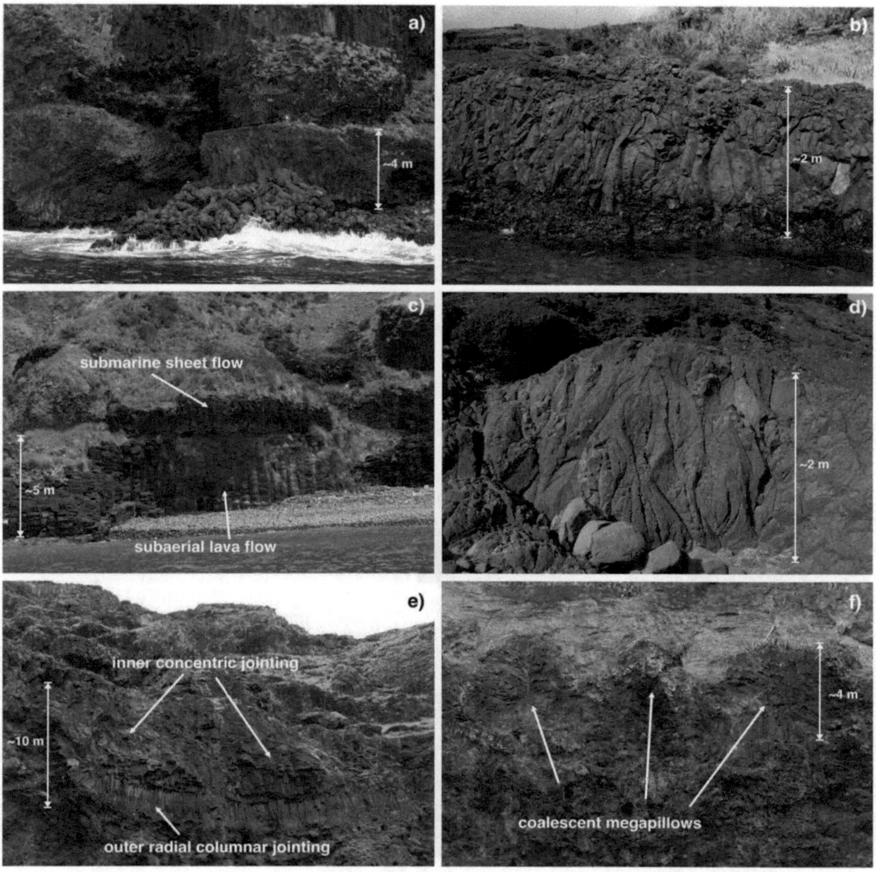

Fig. 4.3 Examples of submarine sheet flows and megapillows from Santa Maria (Azores) and São Nicolau (Cape Verde). **a** Pillow-lavas and sheet flows from Santa Maria. **b** Detail of a sheet flow from Santa Maria, showing the characteristic columnar jointing. **c** Pahoehoe and submarine sheet flows from Santa Maria. Note the difference in the columnar jointing. **d** Submarine sheet flow with flame-like jointing, São Nicolau. **e** Megapillow from Santa Maria. Note the contrast between the inner and outer jointing, possibly due to tube-feeding and inflation processes. **f** Transitional morphology comprising coalescent megapillows, Santa Maria

- rosette-shaped columnar jointing;
- flame-like and fan-like jointing;

As a type example one example from the northern peninsula of Santiago is described in detail (see Fig. 4.4). Here, a 20–30 m thick sheet flow (or flows) is interpreted as submarine, as previous authors suggested [25]. Its present lateral extent may exceed 2 km, attesting to the high flow rate associated with its extrusion. Several features were used to identify this sheet flow as submarine, namely:

Fig. 4.4 Type example of submarine sheet flow from northern Santiago. **a** Sequence in cross section showing submarine sheet flow overlying marine sediments (*foreground*). **b** Detail showing the characteristic columnar jointing characterised by poorly developed, closely packed columns with glassy undulated or crenulated faces. **c** Detail of the sheet flow passing laterally to pillow-lavas. **d** Top of the submarine sheet flow. Note the preserved morphology. **e** Close-up of the flow's top, showing a glassy surface with jigsaw cooling joints

- the flow overlies marine sediments;
- the flow presents the distinct columnar jointing characterised by poorly developed, closely packed columns with glassy undulated or crenulated faces;
- in places, rosette-shaped and fan-like jointing is present;
- the top of the flow is still preserved in places and exhibits a glassy crust with closely packed jigsaw jointing resembling the outer crust commonly found on pillow-lavas but of larger extent. Rare remains of marine sediments can be found on top of the flow;
- most importantly, the sheet lava changes laterally to well preserved pillow-lavas on its edges, where the flow rate waned.

We believe the formation of sheet lavas is not only associated with high extrusion rates, but is often mediated/facilitated by smooth underlying topographies: the majority of the extensive lavas of this type found by us in the Cape Verdes and in Santa Maria (Azores) are usually subhorizontal or gently dipping.

In places, a whole range of transitional morphologies between thick sheet flows, megapillows and pillow-lavas can be found. Transitional morphologies often exhibit undulating surfaces resulting from coalescent tubular megapillows into a single continuous massive flow. Megapillows frequently exhibit a heterogeneous structure in cross section, characterised by a rim of radial columnar jointing that passes very sharply to concentric cooling joints, and with evidence for internal longitudinal flow (see Fig. 4.3). These features suggest a tube-feeding and inflation process similar to the one described by Appelgate & Embley [35] for the sheet flows of Axial volcano in Juan de Fuca Ridge.

4.5.1.4 Lava Deltas

The generation of lava deltas in an important process in coastal and flank growth of ocean island volcanoes [25, 40–47]. Lava deltas, or lava-fed deltas, are often formed when subaerially erupted lavas reach the water and the water column is deep enough to allow a steep subaqueous slope [22, 47]. The entrance of subaerial lavas flows into the water body generates quenching and fragmentation that, due to gravity-driven processes, results in the accumulation of large quantities of coarse-grained, poorly-sorted, volcaniclastic wedges on the progradational front of the advancing flows, just like a Gilbert-type river delta [22, 26, 40, 47, 48] (see Fig. 4.5). Due to the offshore progradation of lava deltas, the resulting structure is composed of three members [48] (see Figs. 4.5 and 4.6):

- *Basal unit* The basal unit is usually composed of a mix of shallow marine sediments (calcarenites, conglomerates etc) and pebble breccias enveloped in sandy hyaloclastite. Bedding varies in dip and normally mantles the underlying topography.
- *Foreset unit* The foreset unit is constituted by steeply dipping (10°–40°) poorly-stratified foreset beds of hyaloclastites and pillow breccias, frequently palagonitised and interbedded with tube-like or ribbon-like pillow-lavas.

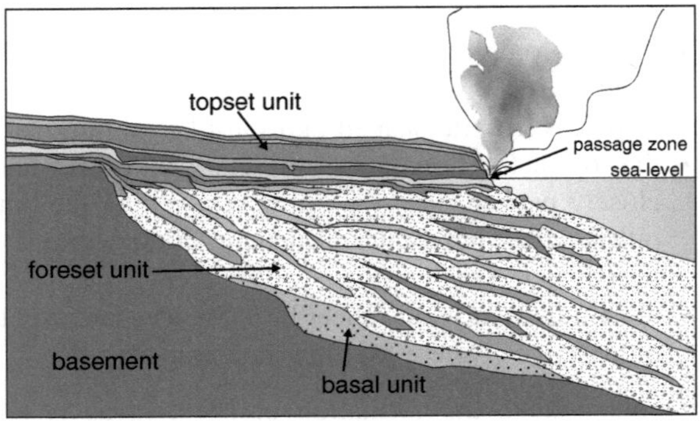

Fig. 4.5 Representation of the structure of a lava delta. Note how the passage zone marks the contemporaneous water level

- *Topset unit* The topset unit is normally composed of a flat-lying columnar-jointed to brecciated submarine sheet lava, or, more frequently of a thick pahoehoe flow or package of pahoehoe flows.

The ratio between the hyaloclastitic material and the pillow-lavas in the foreset unit appears to depend on the flow rate of the system, and possibly also the height of the water column and the underlying topography. If the flow rate is small, fragmentation dominates the water-magma interaction at the front of the delta. However, if the flow rate increases enough to compensate the fragmentation process, pillow-lavas will be generated. Likewise, if the bottom dips gently, the lava flow can more quickly fill the volume between the bottom and the water surface, avoiding breakup and fragmentation and generating pillows. If, like a beach, the sea bottom dips very gently and the flow rate is high, the resultant lava will probably be a submarine sheet flow, as in the 1969–1971 eruption of Kilawea described by Moore et al. [18]. In this case the resultant morphology is not a classic lava delta, and the transition between the subarerial and submarine morphologies is difficult to spot.

In lava deltas, the boundary between the flat-lying lavas (topset) and the steep-dipping breccia and pillow foresets is called the passage zone and can be used to identify unambiguously the water level [22, 40, 48, 49]. As a consequence, coastal lava deltas constitute incontrovertible sea-level markers and are ideal for uplift studies. The work of Jones & Nelson [40] describes the possible structural relationships between temporally successive lava deltas with changes in the water-level either due to eustatic changes or uplift/subsidence. Figure 4.7 illustrates these schematically.

4.5.1.5 Palaeo-depth Estimation Using Submarine Volcanic Morphologies

In order to quantify the vertical displacement undergone by a particular exposed hydromagmatic sequence, one needs to infer as accurately as possible where the sea-level actually was during the extrusion of the sequence. If the sequence is a lava delta and the passage zone is visible, then the passage zone constitutes the

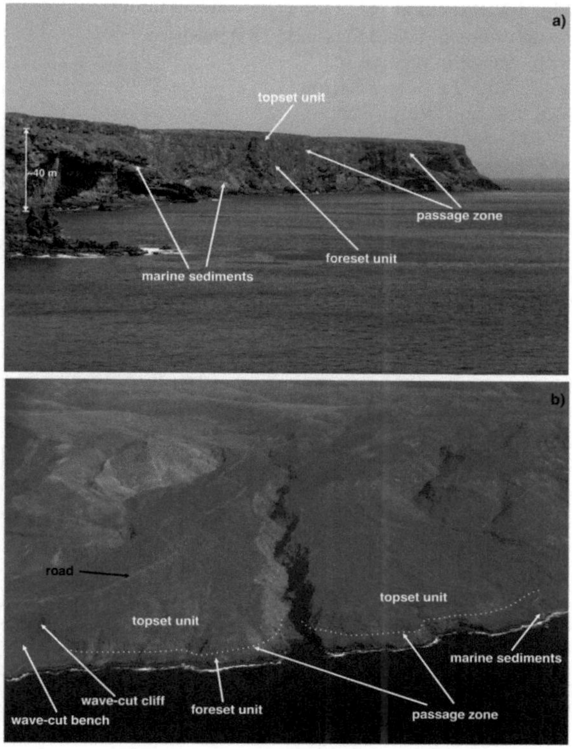

Fig. 4.6 Examples of coastal lava deltas from São Nicolau. **a** Terra Chã lava delta. Note the passage zone between the *topset unit* (subaerial) and the *foreset unit* (submarine), and the *marine sediments* at the base of the delta. **b** Carriçal lava delta. Note the passage zone and the more modern wave-cut morphologies]

marker used against the reference level to infer the vertical displacement. However, many submarine volcanic sequences were subsequently eroded or never reached the sea surface, making a passage zone unavailable for reference. In these situations, unfortunately frequent, one needs to estimate the palaeo-depth of emplacement. The morphometry of vesicles in submarine lavas has been used as a rough indicator of the hydrostatic pressure (hence palaeo-depth) of emplacement, since decreasing pressure increases vesicularity [24, 50, 51]. However, the volatile content and viscosity of magma also affects vesicularity [52], and the weathering of exposed hydromagmatic sequences might erase original features preventing reliable depth reconstructions. Likewise, the presence of explosive volcanic products in a sequence may be used as rough indicator of palaeo-depth. However, the critical depth for explosive volcanism is variable for the same reasons presented above: explosive volcanism occurs when the vapour pressure exceeds the hydrostatic pressure, and the vapour pressure depends on the magma's volatile content [21, 28–30, 52]. Another way to estimate the palaeo-depth of a particular hydromagmatic sequence is to look at fossiliferous sediments when these occur interbedded with the volcaniclastic material. Since fossil associations are frequently characteristic of distinct palaeo-environments, they may help constrain the depth of deposition. In many ocean islands, the only practicable way of estimating

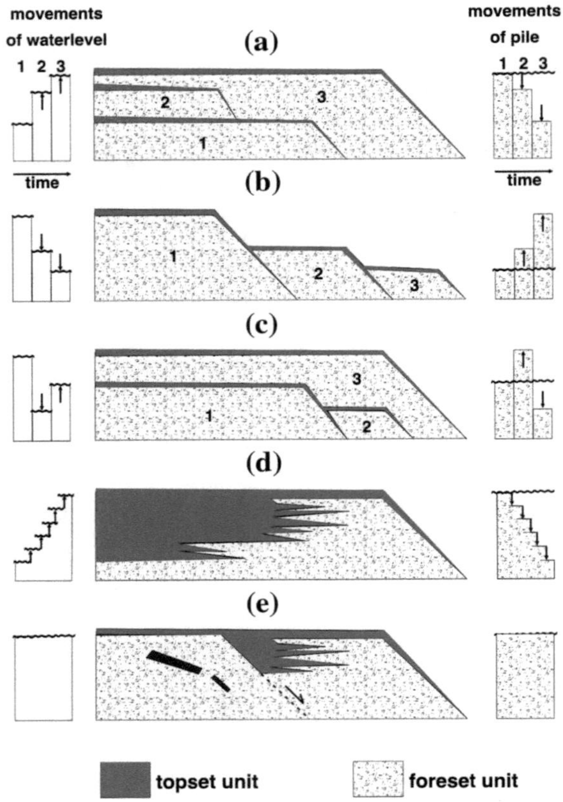

Fig. 4.7 Structural relationships in lava deltas with change in relative sea-level. After Jones and Nelson [40], Cas and Wright [22]

the palaeo-depth of emplacement of their submarine basement complexes is through the fossil content of the interbedded sediments. In the absence of any palaeo-depth indication, one relies instead on the elevation of the highest point at which the submarine sequence crops out as a minimum sea-level height. Figure 4.8 exemplifies the possible situations faced when estimating sea-level height through hydromagmatic sequences.

4.5.2 Marine Sedimentary Formations

The evolution of ocean island volcanoes is a constant battle between constructive and destructive processes [21, 27, 53]. Before emerging, and if the volcanic activity allows, seamounts constitute platforms where sediments accumulate and life thrives. As soon as the edifice emerges, an insular fringing shelf starts to form through a combination of wave erosion and eustatic oscillations [53–57]. The erosional processes that act both on land and on sea eventually generate sediments that provide the necessary substratum for a variety of marine organisms and

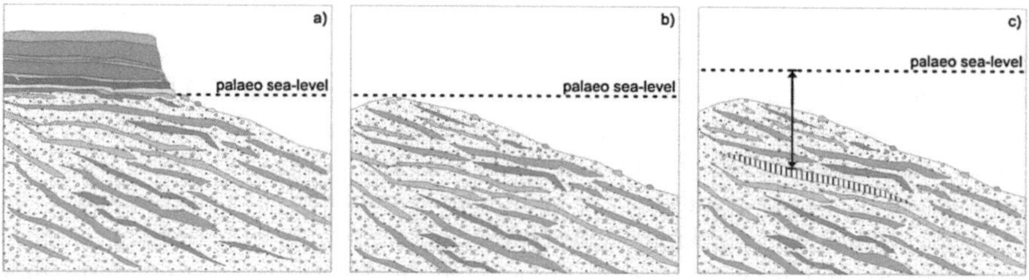

Fig. 4.8 Estimating sea-level height through hydromagmatic sequences. **a** When passage zone is visible, the position of palaeo sea-level corresponds to passage zone. **b** When passage zone is not visible, the position of palaeo sea-level corresponds to the highest point at which the hydromagmatic sequence crops out. In this case the constraint on sea-level a minimum. **c** When passage zone is not visible and marine sediments occur within the hydromagmatic sequence, the position of palaeo sea-level is inferred through the palaeo-depth assigned to the sediments

enhances the biological colonisation of the platform [53, 58–60]. In a similar fashion, as soon as the volcanic activity quiesces (at least locally) the new rocky shores may provide substratum for reef growth if conditions are appropriate. If the volcanic activity resumes, the sediments get incorporated in the island edifice and may later be used to evaluate the environmental conditions in which the volcanic sequence was formed. Hence, marine sedimentary formations constitute important palaeo-markers of sea-level because, due to their fossil content and sedimentological characteristics, one can infer the depositional conditions prevailing during sedimentation and thus estimate the position of sea-level during particular moments of the island history.

The most common marine sedimentary formations found on oceanic islands normally comprise insular shelf sediments, beach sediments and coral reefs (fossil and modern). Occasionally, it is possible to find pelagic or hemipelagic sediments included in the uplifted basement complexes of island edifices. These are very rare but attest to the magnitude of the vertical movements undergone by some seamounts. A general outline of these sedimentary formations follows.

4.5.2.1 Pelagic and Hemipelagic Sediments

Pelagic sediments or pelagites refer to sediments formed in the open sea, comprising more than 75% of biogenic material. The source of the biogenic material is designated as pelagic fallout, and corresponds to the sedimentation of microscopic skeletal remains of calcareous and siliceous planktic organisms that largely live (and die) in the upper part of the water column [60]. Similarly, hemipelagic sediments are formed in rises but are diluted by silt- and clay-sized terrigenous particles [60]. Above the carbon compensation depth (CCD) these types of sediments normally correspond to calcareous oozes, while below this boundary siliceous and clay sediments abound [60].

Like the sea-floor around them, seamounts are covered by the pelagic fallout of biogenic debris. Thus, it is not surprising that pelagic carbonate sediments are found on the flat summits and small flank basins of seamounts [23]. These sediments are normally deposited during pauses in the general volcanic activity, or in areas away from active vents and their hot and acidic products. With the resumption of eruptive activity, the sediments eventually get incorporated in the volcanic edifice. Since pelagic faunas are characteristic of their distinct deep-sea environment, they may help constrain the depth of deposition. Consequently, their presence in uplifted seamount series provides a rare opportunity to more accurately quantify the amount of uplift affecting the sequence. One such example concerns the island of Sal, where pelagic sediments within the basement complex allowed the estimation of the early vertical movements affecting this edifice.

4.5.2.2 Coral Reefs

During and after their erosional stage, many ocean islands often become surrounded by a peripheral reef [26, 61]. When the volcanic activity ceases, and if conditions are appropriate (e.g. the edifice is situated at equatorial latitudes), coral growth becomes increasingly important. A ring-shaped atoll is thus formed, resulting from the continuous upward growth of coral reefs synchronous with island subsidence [27, 62, 63]. When and if island subsidence outpaces reef growth, the atoll disappears beneath the waves and a guyot is them formed [27, 53, 61, 63]. This moment in the atoll's history is called the Darwin point [63]. In contrast, if the island edifice experiences uplift, then the reef structure is exposed to subaerial erosion and a karst morphology is formed. Scott & Rotondo [61] described a whole range of intermediate structures driven by different uplift/subsidence histories, and effectively showing that the study of reef structures may provide useful constraints on island freeboard. Even after drowning, the vertical distance between the top of the guyot and the height of the contemporaneous sea-level may be used to infer a crude subsidence rate e.g. [61, 63]. Hence, the study of coral reefs can both serve to gauge island freeboard e.g. [53, 61, 64, 65]; and to base eustatic studies on e.g. [15, 16, 66].

Judging by the sedimentary formations present in the stratigraphic sequence of the Cape Verde Islands, the development of coral reefs was always incipient and minor. Effectively, and according to Spalding et al. [67], "true reefs" do not occur in Cape Verde due to several factors. One of the factors concerns the existence of a warm, low salinity seasonal current from the Gulf of Guinea, resulting from high riverine discharge, principally from the Niger river, that extends all along the coast from Angola to Mauritania. Outside of this region waters are generally much colder due to upwelling along offshore Western African. Thus, these conditions restrict significant coral reef growth in all of West Africa to shallow protected bays [67]. In the case of Cape Verde, the ongoing volcanic activity, with its hot and acidic products, may also help to prevent coral growth.

4.5.2.3 Shelf and Coastal Sediments

Marine erosion, together with explosive volcanism and stream discharge from the exposed edifice, generates a shelf sediment cover that provides the necessary substratum for a variety of marine organisms [53, 60]. If the insular shelf is wide enough, or if reef structures provide the necessary protection, sediments accumulate on the insular shelf. In a similar fashion, coastal processes dominated by wave erosion form sandy and boulder beaches on the periphery of the volcanic edifice. Biological productivity contributes skeletal remains of shelly mollusks, sea-urchins and other marine organisms, enlarging the sediment availability and leading to an increase in the biogenic component of the sediments. Since biological productivity generally declines with depth, especially for benthic and malacogical faunas, beach and shallow marine sediments are thus richer in whole shells and shell debris [60, 68].

Insular shelves can normally be described, in sedimentological terms, as wave-dominated small open shelves. The sediments that comprise the shelf cover, from the shelf break to the beach, vary greatly in size and distribution according to the prevailing energetic conditions, slope angle, and eustatic variations. Meter-sized boulders to sub-milimeter sands can thus be found, sometimes mixed. As a consequence, when these sediments get incorporated in the island edifice by subsequent eruptions, the resulting consolidated formations vary from sandstones and calcarenites (if the biological productivity was high) to conglomerates.

In old oceanic volcanoes situated at high latitudes or in oceanographic conditions that do not favor coral reef growth, wave erosion is greater since it lacks reef protection. Hence, in these edifices, more extensive marine-cut platforms may form and eventually create the necessary shallow water environments for sand to deposit and for benthic fauna to thrive. Consequently, the generation of sandy shores with a high biogenic content occurs.

Occasionally, if the conditions are propitious, the shelf edge is rimmed by either reef or sand shoals with distinct windward vs leeward characteristics [53, 60]. Windward shelves are usually more extensive due to the higher erosion rates they are subjected to, while on leeward shelves sedimentation is facilitated and complex structures may form by the action of refraction currents. Old and flat-topped volcanic edifices tend to drown rapidly during a major relative sea-level rise, leading to a capping of open marine deeper-water facies or grainstone shoals [60].

4.5.2.4 Palaeo-depth Estimation Using Marine Sedimentary Formations

Marine sediments may provide helpful constraints on the depths at which they were formed through their sedimentological characteristics [68, 69]. The work of Immenhauser [69] presents a recent and complete review of how to interpret palaeo-water depths from the physical rock record.

In a similar manner, the fossil content can be used to constrain depth of deposition of a particular formation [68]. Each environment or sub-environment is usually characterised by a given fauna, i.e., a biocenosis inhabiting the region and called biofacies in palaeontological terms [68]. A particular biocenosis, upon death, burial and preservation, leaves behind a death assemblage (taphocoenosis) that is normally composed of hard skeletal and biogenic parts (such as shells, teeth, etc), sedimental bioturbation structures, bioerosion structures, etc. [68]. This death assemblage, when present in the geological record, may help to reconstruct the biocenosis that is its origin, and thus help to infer past environmental conditions. According to Reineck and Singh [68], benthic communities are most important in characterizing a given environment, because the hard skeletal parts that they leave behind are well preserved in the fossil record. If no significant post-mortem transport occurred, benthic taphocoenoses provide good biofacies characteristic fossils [68]. Macrobenthonic fossils are almost ubiquitous in shallow marine sediments, one group of which is the echinoderm group that are particularly good palaeo-environemnt indicators [70].

Precise palaeo-water depth estimation using shelf and coastal sediments is a difficult task because it involves the determination of precise ecological conditions associated with each fossil species found and this information is, unfortunately, scattered in the literature and in many cases uncertain. In addition, precise species identification and the determination of their ecological niche is beyond the scope of this work. Thus, in order to estimate the palaeo-depth of the sedimentary formations present in the Cape Verde, we used a simplistic assessment based on the general sedimentological features and general fossil assemblage present in the deposits. The existence of a few sedimentary deposits overlain by lava deltas allowed us to calibrate the criteria used for palaeo-water estimates because lava deltas provide evidence regarding the sea-level height contemporaneous with the topmost layers of the sediments.

Table 4.2 summarises the criteria used for palaeo-water depth interpretations. The formulated criteria use the distribution of facies along a typical wave-dominated sandy beach profile (see Fig. 4.9), and along a rocky coast. For more detail refer to Reineck and Singh [68] or Wright and Burchette [60]. Figure 4.10 shows an example of how palaeo-water depth is inferred from a typical marine terrace like the ones found in Sal, Boa Vista or Maio. Figs. 4.11, 4.12 and 4.13 show examples of the deposits described in Table 4.2.

4.5.3 Marine Erosional Morphologies

Marine erosional features are easily recognisable morphologies provided that there was enough time for erosion to act [10, 71], or if the substratum poses a low resistance to erosion. Since marine abrasion is most effective in the breaker zone, due to the combined action of wave and tides [55, 72], its action is stronger in the horizontal direction, effectively contributing to the levelling of reliefs. However, eustatic variations and/or tectonic movements introduce a vertical component that varies according to the amplitude and frequency of the eustatic variations and the magnitude and rate of tectonic uplift/subsidence [55, 72, 73]. Thus, the resulting

Table 4.2 Palaeo-water depth interpretations according to type of deposit

Type of deposit	Main characteristics	Energy level	Palaeo-water depth
Dune/backbeach	Fine grained, very well-sorted and well-rounded calcarenites structured in planar and through crossbeds and megaripples; incipient dunes and laminated bedding in backshore areas	Medium	Above sea-level
Foreshore/shoreface	Fine grained calcarenites in even laminated structures (foreshore deposits); wavy or ripple laminated beds (shoreface); presence of lenses of shell debris and large and thick unarticulated shells	High	0–10 m
Transition zone/offshore	Fine grained unbedded calcarenites with delicate whole organisms like pectinids, branchy bryozoan, echinoderms, etc, many in life position; rich in bioturbation	Low	10–100 m
Boulder beach	Boulder accumulations	Very high	0–5 m
Pebbly beach	Consolidated or unconsolidated conglomerates with large shell debris; presence of thick-shelled mollusks, solitary corals, barnacles, limpets, rhodoliths etc	High	0–10 m
Incipient coral reef	In situ corals attached to rocky substratum and surrounded by coarse grained matrix; presence of pisolitic structures, rhodoliths and shell debris	High	0–15 m
Coralline algae reef	Massive framework of rhodoliths enveloped in a calcarenitic matrix; element or matrix supported; echinoderms may occur	Medium	10–50 m

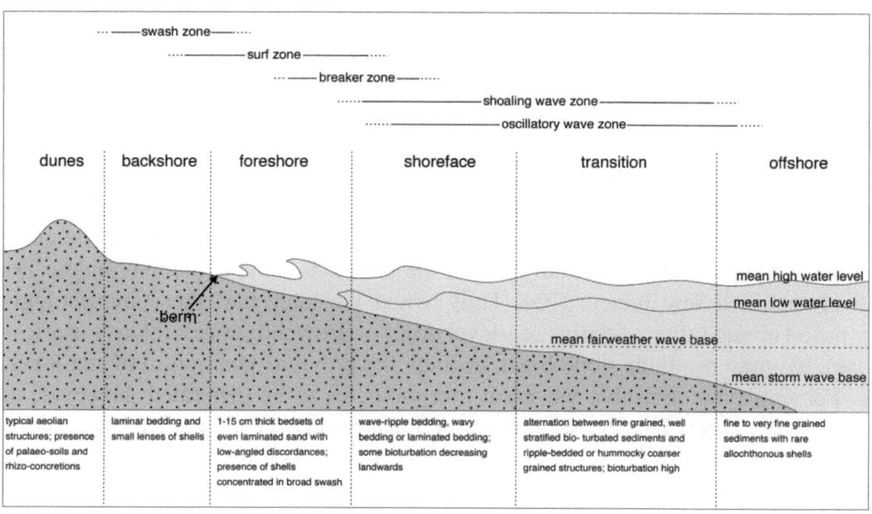

Fig. 4.9 Representation of a typical beach profile, with facies distribution. After Wright and Burchette [60]

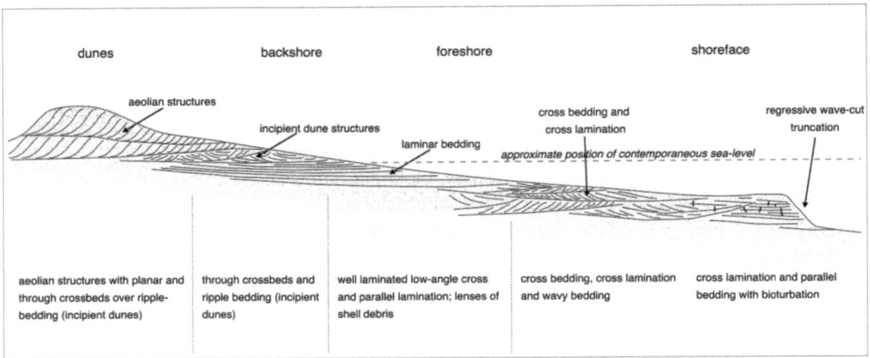

Fig. 4.10 Schematic profile of a typical calcarenite terrace from Sal, Boa Vista or Maio. This schematic profile outlines the main characteristics present in the uplifted terraces from these islands. Note the position of the inferred sea-level

morphology may be complex, but its study is crucial to reconstruct past sea-level positions (and consequently vertical movements).

In order to track the history of relative sea-level positions through marine abrasion morphologies, two different strategies may be adopted: (1) the study of exposed and/or drowned marine-cut (or wave-cut) surfaces, and (2) the study of the insular shelf. These two approaches are in fact complementary, as will be seen.

4.5.3.1 Wave-Cut Surfaces

Wave-cut surfaces, also called wave-cut platforms or wave-cut benches, are narrow gentle-slopping areas generated by marine abrasion on rocky shores [10, 74]. According to Bradley [74] wave attack creates and drives landward a sea cliff, leaving under shallow water a wave-cut bench thinly covered by sediments (Fig. 4.14 exemplifies the morphological features associated with wave erosion on rocky shores). Since marine erosion only acts effectively in very shallow depths, the slope break (called the shore angle) between the steep wave-cut cliff and the gently dipping wave-cut bench actually marks the position of the contemporaneous sea-level within a few meters [75]. Thus, when we later find these morphologies exposed inland (after a relative sea-level drop) or drowned (after a relative sea-level rise), we may use their slope break to track the previous position(s) of sea-level.

For easily recognisable wave-cut morphologies to be generated, however, the relative sea-level must be stationary for long enough [10, 71]. This is achieved when either the eustatic level is at a stand-still, or when the islands are uplifting/subsiding at compatible (but not equal) rates with the eustatic rises/falls [10]. In fact, for extensive wave-cut platforms to be developed, a relative sea-level rise is required, to allow the advancement of wave-cut action inland. Otherwise marine erosion loses efficiency due to the accumulation of sediments in the break zone

Fig. 4.11 Examples of coastal calcarenitic deposits from Cape Verde. **a** Foreshore fine grained calcarenite from a Pleistocene terrace of Boa Vista. **b** Foreshore deposit from a Pleistocene terrace of Sal. Note the gravel-rich bedsets. **c** Shoreface calcarenitic deposit from a Pleistocene terrace of Maio. Note the cross lamination of the deposit, with laminae rich in shell debris. **d** Pliocene offshore calcarenitic deposit from São Nicolau. Deposit is fine grained, unbedded and rich in delicate pectinids, branchy bryozoa and echinoderm spicules. **e, f** Pliocene transition zone calcarenites from Santiago. Note the planar bedding and intense bioturbation of the deposit

[55, 71–76]. The optimum periods for the development of these structures are, thus, when relative sea-level rise is similar to, or slightly higher, than land uplift rates [75]. Consequently, the formation of extensive terraces is enhanced during eustatic high and lowstands when the rates of sea-level fall and rise approximate uplift rates [75] (see Fig. 4.15). Frequently, on ocean islands with steep slopes, if the relative sea-level is not stationary for long enough to produce a broader surface, the only feature marking the position of sea-level at a particular time is a number of aligned wave-cut notches at similar elevations (see Fig. 4.16).

Fig. 4.12 Examples of pebbly beach deposits from Cape Verde. **a** Pleistocene conglomeratic beach deposit from Santiago. **b**, **c** Pleistocene unconsolidated conglomeratic beach deposits from São Nicolau. Note the high energy fossil assemblage. **d** Pleistocene consolidated micro-conglomerate from São Nicolau. The deposit is almost exclusively composed of bioclasts and whole large gastropods. **e** Pleistocene consolidated micro-conglomerate from São Nicolau. Another example with high energy fossil assemblage. **f** Pleistocene consolidated micro-conglomerate from São vicente

In contrast, if the relative sea-level is approximately stationary for long enough, wave-cut action may completely truncate the (old) edifice and raze residual relief [53]. In this situation the guyot stage is thus reached, and the edifice is characterised by a distinctive flat topped morphology [53, 54]. However, if an edifice experiences an uplift trend, the existing platforms are eventually preserved as terraces. This is particularly true for the latest Quaternary terraces. Effectively, according to Blanchon and Shaw [66], Quaternary sea level oscillations are best preserved along shorelines which are subjected to uplift. These authors also suggest that relatively stable coastlines are not conducive to the preservation and

Fig. 4.13 Examples of reef deposits from Cape Verde. **a** Pliocene incipient coral reef deposit from Santiago, between pillow-lavas. **b** Close-up of the previous deposit. Note the coral attachment to the substratum, providing protection for the sedimentation of the deposit. **c** Close-up of deposit (**a**), showing a cluster of rhodoliths. **d** Close-up of deposit (**a**), showing a rhodolith floating in a bioclastic and pisolitic matrix. **e** Pliocene rhodolith reef from São Nicolau. Note the massive accumulation of rhodoliths. Large irregular echinoderms can also be found in the deposit (**f**)

Fig. 4.14 Representation of a wave-cut surface on a rocky shore

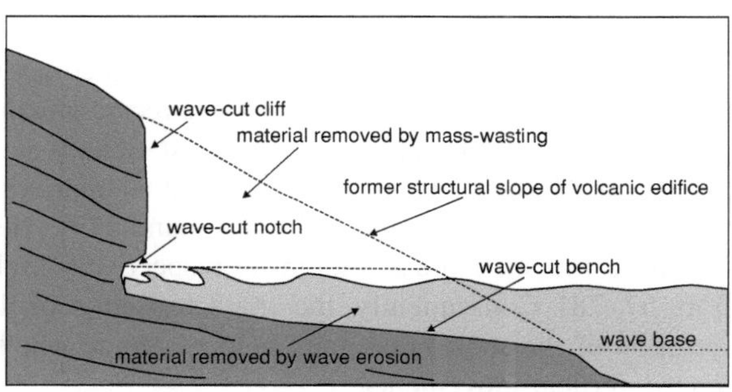

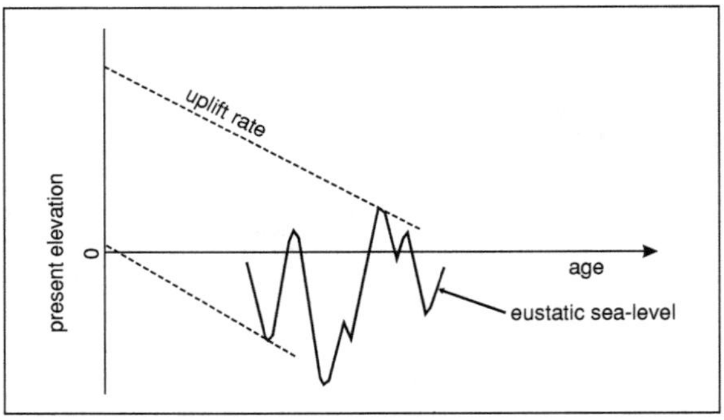

Fig. 4.15 Short periods favourable to the genesis of extensive wave-cut surfaces, in the context of eustatic variations. The sea is capable of creating extensive wave-cut platforms when the rate of sea-level rise is similar or slightly higher to land uplift rates. In the graph, a particular uplift rate is represented by the two uplift rate lines, to show that uplift and eustatic rates become similar near the tangent points. After Yeats et al. [75]; Cabral [77]

subsequent recognition of geomorphic landforms. Thus, the structures that are most helpful for estimating past sea-levels are typically beaches and erosional marine-cut terrace platforms [66]. The older and razed Cape Verdean islands of Sal, Boa Vista and Maio are particularly good case studies of this situation.

4.5.3.2 Insular Shelf Morphology

Strictly speaking, the insular shelf and shelf break are not palaeo-markers of sea-level sensu stricto. However, since their formation is directly connected to a relative sea-level change, one can use it to test if island uplift or subsidence occurred.

The insular shelf is the zone encircling the island edifice from the line of permanent immersion to a point of increasing slope gradient towards the great ocean depths, called the shelf break [53, 54, 78]. The insular shelf is thought to be formed by the combined effects of wave erosion and recent eustatic variations [53]. Since tectonic movements may also affect island edifices, the shelf morphology will depend greatly on the amplitude and frequency of recent eustatic oscillations and the magnitude and rate of tectonic uplift/subsidence [55, 72, 73]. Thus, the insular shelf depends on those two components (eustatic and tectonic), plus the age of the volcanic edifice since recent volcanic eruptions may "reset" the erosive process [53, 57]. If an island is stationary and no recent volcanic eruption has occurred, the shelf break should be at an approximate depth of 130 m [56, 78] which is the value of the LGM lowstand [79]. Hence, if uplift or subsidence occurred the shelf break should be respectively higher or lower than this value [56, 57, 78]. Consequently, the characterisation of the shelf morphology and the position of the shelf break may help to decide whether the island edifice experienced vertical movements.

Fig. 4.16 Examples of wave-cut morphologies from the Azores and Cape Verde. **a** Pleistocene wave-cut bench covered in slope deposit, from Santa Maria; **b** Detail from previous example. Note the wave-cut morphologies. **c** Pleistocene wave-cut notches from Santa Maria; **d** Wave-cut bench and cliff, and boulder beach at 80–100 m asl, Santiago; **e** Wave-cut morphologies and terraces from Boa Vista; **f** Several generations of wave-cut morphologies from São Nicolau

4.6 Summary

One of the major aims of this thesis is to constrain the history of vertical movements affecting the Cape Verde Archipelago, in an attempt to discriminate between competing swell models. In order to do so, we started by defining the criteria and the method on which our vertical movement reconstructions would be based. The chosen method uses sea-level as a reference datum and markers of past relative sea-level as uplift/subsidence benchmarks. The choice was based on the fact that sea-level related markers are frequent in the Cape Verde Islands and allow long-term reconstructions. Subsequently, the principles lying behind the estimation of the vertical displacement experienced by the markers were presented. A simple model was chosen that involves the determination of the present elevation, the palaeo-water depth and the height of the contemporaneous sea-level from the eustatic curve chosen. Then, the criteria used to recognise the variety of palaeo-markers of sea-level (the particular tracers of uplift and subsidence) were outlined and several examples from different archipelagos were presented. The most common palaeo-markers found on ocean islands (and in the Cape Verdes in particular) are of sub-marine volcanic, marine sedimentary, and wave-cut morphological nature.

Submarine volcano palaeo sea-level markers constitute some of the most common and easily recognisable of all markers in Cape Verde. They include pillow-lavas and hyaloclastites, submarine sheet flows (a morphology that is poorly described in the literature) and lava deltas. The former are particularly useful because their passage zone marks unambiguously the contemporaneous sea-level. When the passage zone is not visible, we use the elevation of the highest outcrop of the unit as the minimum contemporaneous sea-level height.

Marine sedimentary units are also frequent and may also help to constrain the position of relative sea-level through the island's geological history. An outline of the main marine sedimentary formations found on ocean islands in general, and in the Cape Verdes in particular, was presented focusing on how palaeo-water depth estimates were made. A number of characteristic deposits were thus presented to facilitate the task of estimating palaeo-water depths.

Wave-cut related morphologies may also help to constrain relative sea-level change, and have a direct link with uplift movements. They are particularly useful in the estimation of recent (Quaternary) movements affecting the archipelago. However, for this work observations were limited to the exposed part of the edifices, i.e. the bathymetry was not investigated in search of drowned wave-cut surfaces or to determine the position of the shelf break. The reason for this is the fact that access to detailed bathymetry of the archipelago was not possible.

References

1. Brookfield, M. (2004). *Principles of stratigraphy*. Oxford, UK: Blackwell.
2. Donovan, D., Jones, E., Ridd, M., & Hubbard, J. (1979). Causes of world-wide changes in sea level; with discussion. *Journal of the Geological Society, 136*(2), 187–193.
3. Haq, B., Hardenbol, J., & Vail, P. (1987). Chronology of fluctuating sea levels since the Triassic. *Science, 235*(4793), 1156–1167.
4. Miller, K., Kominz, M., Browning, J., Wright, J., Mountain, G., Katz, M., et al. (2005). The phanerozoic record of global sea-level change. *Science, 310*(5752), 1293–1298.
5. Bamler, R., & Hartl, P. (1998). Synthetic aperture radar interferometry. *Inverse Problems, 14*(4), R1–R54.
6. Massonnet, D., & Feigl, K. (1998). Radar interferometry and its application to changes in the Earth's surface. *Reviews of Geophysics, 36*(4), 441–500.
7. Burgmann, R., Rosen, P., & Fielding, E. (2000). Synthetic aperture radar interferometry to measure Earth's surface topography and its deformation. *Annual Review of Earth and Planetary Sciences, 28*(1), 169–209.
8. Suess, E., Sollas, W., & Sollas, H. (1906). *The face of the earth*. Oxford, UK: Clarendon.

9. Reading, H. (1996). *Sedimentary environments: processes, facies and stratigraphy* (3rd ed.). Oxford, UK: Blackwell.
10. Bird, E. (2008). *Coastal geomorphology: an introduction* (2nd ed.). Chichester, UK: Wiley.
11. Posamentier, H., Jervey, M., & Vail, P. (1988). Eustatic controls on clastic deposition II—sequence and system tract models. In C. Wilgus, B. Hastings, H. Posamentier, J. Van Wagoner, C. A. Ross, & C. Kendall (Eds.), *Sea level changes: an integrated approach* (vol. 42, pp. 125–154). Society of Economic Palaeontologists and Mineralogists Special Publication.
12. Posamentier, H., & James, D. (1993). An overview of sequence-stratigraphic concepts: uses and abuses. In C. Summerhayes, B. Haq, & G. Allen (Eds.), *Sequence stratigraphy and facies associations* (vol. 18, pp. 3–18). International Association of Sedimentologists Special Publication.
13. Shackleton, N. (1987). Oxygen isotopes, ice volume and sea level. *Quaternary Science Reviews, 6*(3–4), 183–190.
14. Miller, K., Fairbanks, R., & Mountain, G. (1987). Tertiary oxygen isotope synthesis, sea level history, and continental margin erosion. *Paleoceanography, 2*(1), 1–19.
15. Chappell, J., Omura, A., Esat, T., McCulloch, M., Pandolfi, J., Ota, Y., & Pillans, B. (1996). Reconciliaion of late quaternary sea levels derived from coral terraces at Huon Peninsula with deep sea oxygen isotope records. *Earth and Planetary Science Letters, 141*(1–4), 227–236.
16. Yokoyama, Y., Esat, T., & Lambeck, K. (2001). Coupled climate and sea-level changes deduced from Huon Peninsula coral terraces of the last ice age. *Earth and Planetary Science Letters, 193*(3–4), 579–587.
17. Torres, P., Silva, L., Serralheiro, A., Tassinari, C., & Munhá, J. (2002). Enquadramento geocronológico pelo método K/Ar das principais sequências vulcano-estratigráficas da Ilha do Sal—Cabo Verde. *Garcia de Orta, Serviços Geológicos, 18*(1–2), 9–13.
18. Moore, J., Phillips, R., Grigg, R., Peterson, D., & Swanson, D. (1973). Flow of lava into the sea, 1969-1971, Kilauea Volcano, Hawaii. *Bulletin of the Geological Society of America, 84*(2), 537–546.
19. Moore, J., Clague, D., & Normark, W. (1982). Diverse basalt types from Loihi seamount, Hawaii. *Geology, 10*(2), 88–92.
20. Moore, J., & Clague, D. (1987). Coastal lava flows from Mauna Loa and Hualalai volcanoes, Kona, Hawaii. *Bulletin of Volcanology, 49*(6), 752–764.
21. Schmincke, H.-U. (2004). *Volcanism* (1st ed.). Berlin: Springer.
22. Cas, R., & Wright, J. (1987). *Volcanic successions. modern and ancient: a geological approach to processes, products and successions*. London, UK: Chapman & Hall.
23. Batiza, R., & White, J. (2000). Submarine lavas and hyaloclastite. In H. Sigurdsson, B. Houghton, S. McNutt, H. Rymer, & J. Stix (Eds.), *Encyclopedia of volcanoes* (pp. 361–382). Salt Lake City, UT: Academic Press.
24. Moore, J., & Schilling, J. (1973). Vesicles, water, and sulfur in Reykjanes Ridge basalts. *Contributions to Mineralogy and Petrology, 41*(2), 105–118.
25. Serralheiro, A. (1976). A Geologia da Ilha de Santiago (Cabo Verde). *Boletim do Museu e Laboratorio Mineralógico e Geológico da Faculdade de Ciências, 14*, 157–369.
26. Orton, G. (1996). Volcanic environments. In H. Reading (Ed.), *Sedimentary environments: processes, facies and stratigraphy* (pp. 485–567). Oxford, UK: Blackwell.
27. Schmidt, R., & Schmincke, H.-U. (2000). Seamounts and island building. In H. Sigurdsson, B. Houghton, S. McNutt, H. Rymer, & J. Stix (Eds.), *Encyclopedia of volcanoes* (pp. 383–402). Salt Lake City, UT: Academic Press.
28. Sourirajan, S., & Kennedy, G. (1962). The system H^2O–NaCl at elevated temperatures and pressures. *American Journal of Science, 260*(2), 115.
29. McBirney, A. (1963). Factors governing the nature of submarine volcanism. *Bulletin of Volcanology, 26*(1), 455–469.
30. Fisher, R. (1984). Submarine volcaniclastic rocks. *Geological Society of London Special Publications, 16*(1), 5–27.

31. Macedo, J., Serralheiro, A., & Silva, L. (1988). Notícia Explicativa da Carta Geológica da Ilha de S Nicolau (Cabo Verde) na escala de 1:50000, Garcia de Orta. *Serviços Geológicos, 11*(1–2), 1–32.
32. Holcomb, R., Moore, J., Lipman, P., & Belderson, R. (1988). Voluminous submarine lava flows from Hawaiian volcanoes. *Geology, 16*(5), 400–404.
33. Binard, N., Stoffers, P., Hekinian, R., & Cheminee, J. (2004). South Pacific intraplate volcanism: structure, morphology, and style of eruption. In R. Hekinian, P. Stoffers, & J. Cheminee (Eds.), *Oceanic hotspots: intraplate submarine magmatism and tectonics* (pp. 157–207). Berlin: Springer.
34. Ballard, R., Holcomb, R., & van Andel, T. (1979). The Galapagos Rift at 86 W: 3. Sheet flows, collapse pits, and lava lakes of the rift valley. *Journal of Geophysical Research (Solid Earth), 84*(B10), 5407–5422.
35. Appelgate, B., & Embley, R. (1992). Submarine tumuli and inflated tube-fed lava flows on Axial Volcano, Juan de Fuca Ridge. *Bulletin of Volcanology, 54*(6), 447–458.
36. Sigvaldason, G. (1968). Structure and products of subaquatic volcanoes in Iceland. *Contributions to Mineralogy and Petrology, 18*(1), 1–16.
37. Saemundsson, K. (1970). Interglacial lava flows in the lowlands of southern Iceland and the problem of two-tiered columnar jointing. *Jokull, 20*, 62–77.
38. Long, P., & Wood, B. (1986). Structures, textures, and cooling histories of Columbia River basalt flows. *Bulletin of the Geological Society of America, 97*(9), 1144–1155.
39. Bergh, S., & Sigvaldason, G. (1991). Pleistocene mass-flow deposits of basaltic hyaloclastite on a shallow submarine shelf, South Iceland. *Bulletin of Volcanology, 53*(8), 597–611.
40. Jones, J., & Nelson, P. (1970). The flow of basalt lava from air into water, its structural expression and stratigraphic significance. *Geological Magazine, 107*(1), 13–19.
41. Jones, J., & McDougall, I. (1973). Geological history of Norfolk and Philip Islands, southwest Pacific Ocean. *Australian Journal of Earth Sciences, 20*(3), 239–254.
42. Furnes, H., & Sturt, B. (1976). Beach/shallow marine hyaloclastite deposits and their geological significance: an example from Gran Canaria. *The Journal of Geology, 84*(4), 439–453.
43. Peterson, D. (1976). Processes of volcanic island growth, Kilauea Volcano, Hawaii, 1969–1973. In *Proceedings of the Symposium on Andean and Antarctic Volcanology Problems Specification Series International Association* (pp. 172–189).
44. Lipman, P., & Moore, J. (1996). Mauna Loa lava accumulation rates at the Hilo drill site: Formation of lava deltas during a period of declining overall volcanic growth. *Journal of Geophysical Research (Solid Earth), 101*(B5), 11631–11641.
45. Schmincke, H., Behncke, B., Grasso, M., & Raffi, S. (1997). Evolution of the northwestern Iblean Mountains, Sicily: uplift, Plicocene/Pleistocene sea-level changes, paleoenvironment, and volcanism. *Geologische Rundschau, 86*(3), 637–669.
46. Smith, J., Malahoff, A., & Shor, A. (1999). Submarine geology of the Hilina slump and morpho-structural evolution of Kilauea volcano, Hawaii. *Journal of Volcanology and Geothermal Research, 94*(1–4), 59–88.
47. Skilling, I., White, J., & McPhie, J. (2002). Peperite: a review of magma–sediment mingling. *Journal of Volcanology and Geothermal Research, 114*(1–2), 1–17.
48. Porebski, S., & Gradzinski, R. (1990). Lava-fed Gilbert-type delta in the Polonez Cove Formation (Lower Oligocene), King George Island, West Antarctica. In A. Colella, & D. Prior (Eds.), *Coarse Grained Deltas* (vol. 10). International Association of Sedimentologists Special Publication (pp. 335–351).
49. Smellie, J. (2000). Subglacial eruptions. In H. Sigurdsson, B. Houghton, S. McNutt, H. Rymer, & J. Stix (Eds.), *Encyclopedia of volcanoes* (pp. 403–418). Salt Lake City, UT: Academic Press.
50. Jones, J. G. (1969). Pillow lavas as depth indicators. *American Journal of Science, 267*(2), 181–195.
51. Dixon, J., Stolper, E., & Holloway, J. (1995). An experimental study of water and carbon dioxide solubilities in mid-ocean ridge basaltic liquids. Part I: Calibration and Solubility Models. *Journal of Petrology, 36*(6), 1607–1631.

52. Sparks, R. (1978). The dynamics of bubble formation and growth in magmas: A review and analysis. *Journal of Volcanology and Geothermal Research, 3*(1–2), 1–37.
53. Menard, H. (1983). Insular erosion, isostasy, and subsidence. *Science, 220*(4600), 913–918.
54. Dietz, R., & Menard, H. (1951). Origin of abrupt change in slope at continental shelf margin. *AAPG Bulletin, 35*(9), 1994–2016.
55. Trenhaile, A. (2000). Modeling the development of wave-cut shore platforms. *Marine Geology, 166*(1), 163–178.
56. Trenhaile, A. (2001). Modelling the Quaternary evolution of shore platforms and erosional continental shelves. *Earth Surface Processes and Landforms, 26*(10), 1103–1128.
57. Quartau, R., Trenhaile, A., Mitchell, N., & Tempera, F. (2000). Modelling the development of a volcanic island shelf by erosion: Faial Island in the Azores Arquipelago. *Marine Geology, 275*(1–4), 66–83.
58. Gulick, A. (1932). Biological peculiarities of oceanic islands. *Quarterly Review of Biology, 7*(4), 405–427.
59. Sauer, J. (1969). Oceanic islands and biogeographical theory: a review. *Geographical Review, 59*(4), 582–593.
60. Wright, V., & Burchette, T. (1996). Shallow-water carbonate environments. In H. Reading (Ed.), *Sedimentary environments: processes, facies and stratigraphy* (pp. 325–394). Oxford, UK: Blackwell.
61. Scott, G., & Rotondo, G. (1983). A model to explain the differences between Pacific plate island-atoll types. *Coral Reefs, 1*(3), 139–150.
62. Darwin, C. (1842). *On the structure and distribution of coral reefs*. Ward Lock: London, UK.
63. Grigg, R. (1982). Darwin Point: a threshold for atoll formation. *Coral Reefs, 1*(1), 29–34.
64. Menard, H., & Ladd, H. (1963). Oceanic islands, seamounts, guyots and atolls. *The Sea, 3*, 365–385.
65. McNutt, M., & Menard, H. W. (1978). Lithospheric flexure and uplifted Atolls. *Journal of Geophysical Research (Solid Earth), 83*(B3), 1206–1212.
66. Blanchon, P., & Shaw, J. (1995). Reef drowning during the last deglaciation; evidence for catastrophic sea-level rise and ice-sheet collapse. *Geology, 23*(1), 4–8.
67. Spalding, M., Ravilious, C., & Green, E. (2001). *World atlas of coral reefs*. Berkeley, CA: University of California Press.
68. Reineck, H., & Singh, I. (1980). *Depositional sedimentary environments: with reference to terrigenous clastics*. Berlin: Springer.
69. Immenhauser, A. (2009). Estimating palaeo-water depth from the physical rock record. *Earth Science Reviews, 96*(1–2), 107–139.
70. Kroh, A., & Nebelsick, J. (2003). Echinoid assemblages as a tool for palaeoenvironmental reconstruction-an example from the Early Miocene of Egypt, Palaeogeography, Palaeoclimatology. *Palaeoecology, 201*(1), 157–177.
71. Trenhaile, A. (1987). *The geomorphology of rock coasts*. Oxford, UK: Clarendon Press.
72. Trenhaile, A. (2002). Rock coasts, with particular emphasis on shore platforms. *Geomorphology, 48*(1–3), 7–22.
73. Trenhaile, A. (2004). Modeling the accumulation and dynamics of beaches on shore platforms. *Marine Geology, 206*(1–4), 55–72.
74. Bradley, W. (1958). Submarine abrasion and wave-cut platforms [California]. *Bulletin of the Geological Society of America, 69*(8), 967–974.
75. Yeats, R., Sieh, K., & Allen, C. (1997). *The geology of earthquakes*. New York: Oxford University Press.
76. Trenhaile, A. (2005). Modelling the effect of waves, weathering and beach development on shore platform development. *Earth Surface Processes and Landforms, 30*(5), 613–634.
77. Cabral, J. (1998). *Elementos de Geomorfologia* (1st ed.). Lisbon, Portugal: Associação dos Estudantes da Faculdade de Ciências de Lisboa.
78. Shepard, F. (1973). *Submarine geology*. New York: Harper Collins.
79. Yokoyama, Y., Lambeck, K., De Deckker, P., Johnston, P., & Fifield, L. (2000). Timing of the last glacial maximum from observed sea-level minima. *Nature, 406*(6797), 713–716.

Chapter 5
Tracers of Uplift and Subsidence in the Cape Verde Archipelago

5.1 Introduction

The growth and decay of ocean islands is normally accompanied by vertical movements. These movements probably result from the combination of several processes, chiefly: (1) the ageing of the underlying lithosphere [1]; (2) lithospheric flexural response to the intrinsic mass transfer associated with island volcanism (surface and sub-surface loading) [2–6], and consequent redistribution of volume by erosion and mass-wasting effects [7–9]; (3) hotspot swell development and dynamics [2, 10–13]. By understanding the history of vertical movements in ocean islands one can assess the individual contribution of the various uplift mechanisms, which may help to test different models for the subsurface processes.

The Cape Verde hotspot is unique among major hotspots because it lies on the African plate, which is approximately stationary in the hotspot reference frame [14–16], and the island chain is built on top of the largest bathymetric anomaly in the oceans, the 2 km high Cape Verde rise [17] (see Fig. 5.1).

The islands are grouped in two chains (São Nicolau-Santo Antão and Sal-Brava) in a horseshoe shape [11]. The volcanism responsible for the island building is reported to be older than $\sim$26 Ma [19] and there is no evident hotspot track: volcanism seems to occur in a cluster instead of in a line. Nevertheless, there is the suggestion of an age progression in the southern chain, from east (oldest islands) to west (youngest islands), judging by the geomorphology of the edifices and the age of the oldest exposed lithologies [19–21]. The islands seem to follow an evolutionary history divided into seamount, shield-building, erosional, rejuvenated (or post-erosional) and guyot stages. Due to its stationary position regarding the melting source, the volcanic lifetime of each individual island seems to be longer than the prototypical Hawaiian Islands, as suggested by Schmincke [9] for the Canaries. Likewise, the occurrence of several post-erosional stages (in the older islands) intercalated with long quiescence periods seems also to be a defining characteristic of the Cape Verde island's evolutionary history (see [19, 21–25]).

R. A. S. Ramalho, *Building the Cape Verde Islands*, Springer Theses,

DOI: 10.1007/978-3-642-19103-9_5,

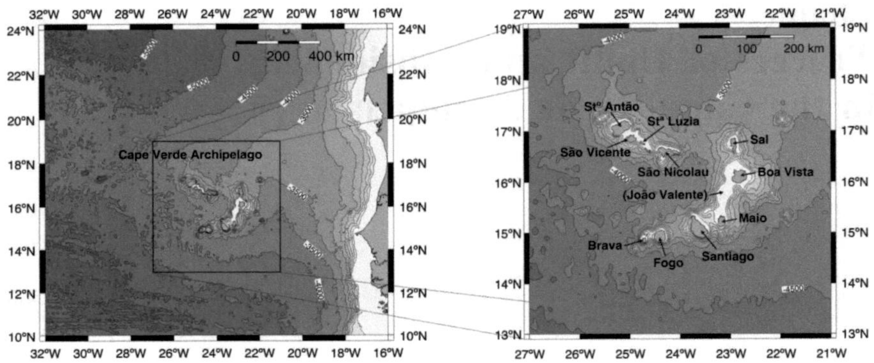

Fig. 5.1 Map of the Cape Verde Archipelago within the Cape Verde bathymetric swell, and detail. Note the existence of a "northern chain" comprising the islands of Santo Antão, São Vicente, Santa Luzia & islets, and São Nicolau, and a "southern chain" comprising Sal, Boa Vista, João Valente (a seamount/guyot), Maio, Santiago, Fogo and Brava. Elevation data from Amante and Eakins [18]

A remarkable feature of the Cape Verde Archipelago is the relative abundance of sea-level palaeo-markers in the islands' stratigraphical sequence, many of which are dateable. By tracking the relative sea-level changes using the age and the present elevation of the markers and comparing them with the eustatic curve, one may infer the vertical movements of the islands.

An overview of the geology and morphology of the islands reveals the existence of two main types of volcanic edifice: prominent (high-relief) and razed (denuded). The prominent volcanoes are rugged with steep slopes and reach higher elevations. The best example of this end-member is Fogo Island: a vigorously active volcano still in its shield-building stage. The islands of Santiago, Santo Antão and São Nicolau are other examples, in the early post-erosional stages [21, 25, 26]. In these islands deep valleys already cut the sequence but the rough dimensions and morphology of the edifices are still preserved. Razed volcanoes, the opposite end-member of the spectrum, are low-lying and flat, frequently covered by extensive marine terraces, and only interrupted by occasional residual relief or small volcanic cones from the late post-erosional stage. Since erosion has already reached deeper levels in these edifices, outcrops of plutonic rocks are common. Good examples of this type are the islands of Sal, Boa Vista and Maio. This spectrum of morphologies is especially evident in the southern chain, where the older and razed volcanoes of the east contrast with the younger and prominent volcanoes of the west.

Since the occurrence of palaeo-markers of sea-level change depends to a great extent on the volcanostratigraphy and the morphology of the islands, different information can be extracted from different islands. The razed volcanoes have lost the majority of their older palaeo-markers to erosion, but exhibit the best record of recent vertical movements due to the prominent marine terraces that cover their eroded surfaces. On the other hand, prominent post-erosional edifices exhibit a longer record because their deep valleys facilitate access to the inner in-sequence formations. In young volcanoes like Fogo, the information we can extract is very poor since the younger lavas completely cover the evidence of past sea-level positions.

In this paper we present a general overview of palaeo-markers of sea-level height that can be found in the Cape Verde Archipelago. We describe key palaeo-markers, especially on the islands of Sal, Santiago and São Nicolau, due to their unique geological record, and interpret their uplift histories for insights into uplift mechanisms. For this work we have analysed the volcanostratigraphic sequence of the islands in the field using the stratigraphy defined by previous authors (when available), complemented with our own field observations and mapping, in order to determine relative elevation changes.

5.2 Definition of Tracers of Uplift or Subsidence

In order to reconstruct the history of vertical movements affecting an ocean island one must both define a reference level and have available a series of dateable past position markers. The reference level is usually the mean sea-level because it provides a global reference (with the relevant isostatic corrections) that is usable world-wide. However, mean sea-level has changed through geological time and consequently any position relative to it depends on the global change in the eustatic level and the island freeboard. Since we have constraints on the former and its uncertainty through existing eustatic curves, we can calculate the island freeboard. For the work reported in this paper we used the eustatic curve by Miller et al. [27], since it is the only recent eustatic curve that extends through the Oligocene, the age of the oldest exposed rocks associated with the Cape Verde volcanism. For each tracer we must determine its present elevation relative to the present sea-level, its age, and (if possible or applicable) its depth of formation.

Any geological structure that can provide us with an estimate of the palaeo-position of sea-level can be used as tracer of island freeboard, provided that it is dateable. The most common sea-level palaeo-markers found on ocean islands are:

- *submarine volcanic units* these comprise volcanic rocks emplaced entirely in a submarine environment or units where the subaerial/submarine transition was eroded or is not observable. These units exclusively exhibit morphologies of submarine eruption such as pillow-lavas, lobate-lavas, hyaloclastite breccia, peperites, etc. Their geological context must be understood, because only units formed in a marine environment can be used. Since the whole unit was formed underwater, its present highest point marks the minimum sea-level height at the time of emplacement. The lack of a solid palaeo-depth estimation method for submarine volcanic formations means that the sea-level height obtained with this kind of marker might be underestimated. These markers are potentially dateable radiometrically.

- *lava deltas* lava deltas are formed when subaerial lava flows reach the sea, generating pillow-lavas and hyaloclastite breccia. In these sequences the boundary between the flat-lying lavas and steep-dipping breccia and pillow foresets—the passage zone—can be used to identify unambiguously the water level [28]. The lavas in (or close to) the passage zone are potentially radiometrically dateable.
- *marine sediments* these are usually marine carbonates and calcarenites, or lithic conglomerates/sandstones deposited in a marine environment. Its fossil association (including ichno-fossils), when present, is diagnostic and a palaeo-depth estimator. Quaternary shallow marine terraces (normally formed by carbonate deposits) are particularly abundant and sometimes extensive in ocean islands and on Cape Verde in particular. These terraces were normally formed during the Quaternary highstands and are frequently associated with marine abrasion surfaces. The top of the terrace, especially if the shore angle is visible, is inferred to be no more than 5 m below the contemporaneous sea-level since wave-cut erosion in rocky shores is concentrated very near the surface [29, 30]. If the sediments include bivalve/gastropods and coral fossils, strontium isotope stratigraphy and U-Th techniques can be used to date its formation.
- *marine abrasion surfaces (wave-cut platforms) and associated structures* these structures are characterised by flat surfaces (sometimes very extensive) with beach deposits (sand and boulder accumulations), carbonate terraces, ichno-fossils (like echinoid holes), etc. The morphology of the contemporaneous backbeach cliff is sometimes still recognisable further inland. The palaeo-depth of these features is assumed to less than 5 m above the top of the surface, for the reasons previously explained. Without fossils or a volcanic cover, these features are extremely difficult to date. Cosmogenic exposure age dating techniques could eventually be used provided that the rocks were never exposed to cosmogenic irradiation prior the formation of the abrasion surface or subsequently re-submerged (creating a complex exposure history).

For each tracer it is possible to calculate the vertical displacement relative to the contemporaneous sea-level provided by the chosen eustatic curve. Thus the vertical displacement of a particular tracer of sea-level height can be expressed as:

$$D_v = h + d - H \qquad (5.1)$$

where D_v is vertical displacement, h is the present elevation, d the inferred palaeo-depth, and H is the contemporaneous sea-level height extracted from the eustatic curve. H is normally expressed as the mean sea-level height for the unit age interval or a particular numerical age uncertainty interval (inferred from published numerical ages). The present elevations were determined using a barometric altimeter, regularly calibrated to avoid diurnal drift, in conjunction with a 1/25,000 topographical map. Errors in the determination of the elevations were neglected, due to their relative small size relative to palaeo-depth estimations or age determinations: we estimate the elevation errors associated with markers closer to a calibration point (like trigonometric stations or the present sea-level) to be around

± 1 m; errors associated with markers distant from calibration points are estimated to be no more than ± 5 m. Likewise the uncertainty in the eustatic curve was also neglected.

5.3 Tracers of Uplift and Subsidence in the Cape Verde Geological Record

5.3.1 Sal

Sal is one of the lowest islands in Cape Verde. It is flat-lying and its only relief is a few recent (<1 Ma) cones and the residual prominence of the older Serra Negra volcano.

The island's volcanostratigraphy was defined by Silva et al. [24] and is summarised in Table 5.1 at the end of the chapter, with age and inferred sea-level constraints. The exposed geology is complex and is the result of events spanning from the Late Oligocene/Early Miocene to the present [19] whose history is still poorly understood. The evolution of the edifice seems to have been characterised by discrete periods of volcanic activity involving increasingly lower volumes of erupted magma, and separated by important periods of erosion lasting up to 3–4 Ma each. We shall overview the evolutionary history of the island, highlighting the evidence for past sea-level positions.

comprises [19, 24]: Ancient Eruptive Complex; Principal Eruptive Formation; Ponta do Altar-Baleia Complex; Conglomerate Breccia; Ribeira da Fragata Formation; Serra Negra Eruptive Formation; Monte Grande-Pedra Lume Formation; and Quaternary Sediments (see Fig. 5.2).

5.3.1.1 The Ancient Eruptive Complex

The oldest unit, the Ancient Eruptive Complex, is profoundly eroded but probably corresponds to the remains of the seamount stage. It occurs mainly in the central part of the island where it reaches the maximum elevation of $\sim$60 m asl. Silva et al. [24] and Torres et al. [31] describe the Ancient Eruptive Complex as a complex sequence of formations mostly comprising hydrovolcanic products of effusive and explosive nature. The whole unit is intensively intruded by dykes—up to 95% of the outcropping area—and small plutonic bodies, a condition that in conjunction with the advanced state of weathering exhibited by its products is an obstacle for solid event reconstructions.

The oldest products seem to comprise weathered basaltic pillow-lavas, pillow-breccias and hyaloclastites [19] probably extruded in the Late Oligocene as the 25.6 ± 2.2 Ma K-Ar age suggests [31], making Sal one of the oldest, if not the oldest, islands in the archipelago. Volcanic breccias of trachytic, trachyphonolitic

Table 5.1 Summarized volcanostrigraphy of Sal, with age and sea-level constraints

Units[a,b]	Main characteristics	Age data for exposed rocks	Evidence for contemporaneous sea-level	Palaeo-depth constraints	Sea-level constraints	Key localities for sea-level markers
Quaternary sediments	Comprise extensive marine terraces (and wave-cut surfaces), dunes, alluvium, colluvium, etc	Quaternary[a,c]	Marine terraces and wave-cut surfaces ranging 1–100 m asl (16–17 terraces)[a,b,c]	Beach deposits and shallow marine sediments. Inferred palaeo-depth <5 m	Sea-level up to 105 m asl	Center-South part of the island, especially from Algodoeiro to Serra Negra
Erosional and conformable contacts, depending on localities						
Monte Grande-Pedra Lume Fm	Post-erosional stage; strombolian-hawaiian and tuff cones with few subaerial flows[b]	0.6 ± 0.4 Ma (K/Ar)[b]; 1.06 ± 0.03 Ma (Ar/Ar)[d]	Unit mostly subaerial, cropping out down to the present sea-level	Phreatomagmatic cone may indicate proximity to sea-level	Sea-level at or slightly below present level	Pedra-Lume; Monte Grande
Erosional contact						
Serra Negra eruptive Fm	Post-erosional stage lava delta[a,b]	5.6 ± 0.4 Ma (K/Ar)[b]; 5.43 ± 0.05 Ma (Ar/Ar)[d]	Passage zone is clearly visible at 90 m asl	Passage zone is clearly visible at 90 m asl	Sea-level at 90 m asl	Serra Negra
Conformable contact						
Ribeira da Fragata Fm	Marine sediments; correspond to volcanic quiescence and erosion[b]	Late Miocene/Pliocene[b]	Marine sediments up to 50 ma asl	5–40	Sea-level up to 55–90 m asl	Ribeira de Fragata; Serra Negra
Erosional contact						
Conglomerate-breccia deposit	Lahar deposits; correspond to volcanic quiescence and erosion[b]	Unknown	Unknown	–	Unknown	–

(continued)

Table 5.1 (continued)

Units[a,b]	Main characteristics	Age data for exposed rocks	Evidence for contemporaneous sea-level	Palaeo-depth constraints	Sea-level constraints	Key localities for sea-level markers
Erosional contact						
Ponta do Altar-Baleia Eruptive Complex	Effusive subaerial and submarine basalts; shallow marine sediments intercalated; partially synchronous with Principal Eruptive Complex[b]	8.7 ± 1.2 Ma (K/Ar)[b]; 11.23 ± 0.09 Ma (Ar/Ar)[d]	Alternation between subaerial and submarine morphologies; marine sediments; submarine facies up to 32 m asl	Alternation between subaerial and submarine facies; proximity to sea-level	Sea-level oscillations around 0–35 m asl	Baleia
Nature of contact not clear						
Principal Eruptive Complex	Effusive subaerial and submarine basalts; intruded by dyke swarm and small pockets of plutonic rocks[b]	Between 15.8 ± 1.8 and 9.7 ± 1 Ma (K/Ar)[b]; 14.2 ± 0.3 Ma (Ar/Ar)[d]	Possible passage zone around 50 m asl; later subaerial lavas down to present level	Possible passage zone around 50 m asl	Sea-level up to 50 m asl	Ribeira do Lizardo
Erosional contact						
Ancient Eruptive Complex	Deeply weathered submarine products; pervasively intruded by dyke swarm and small pockets of plutonic rocks[b]; pelagic carbonates interbedded in the sequence[e]; final products of seamount stage?	25.6 ± 2.2 Ma (K/Ar)[b]; pelagic carbonates suggest Mid Miocene[e,f]	Entire unit is submarine cropping out up to 50 m asl	Pelagic carbonates suggest >300 m depth	Sea-level in excess of 350 m asl	Central region of the island

[a] Silva et al. [24]
[b] Torres et al. [31]
[c] Zazo et al. [36]
[d] Holm et al. [21]
[e] Ubaldo et al. [32]
[f] This study

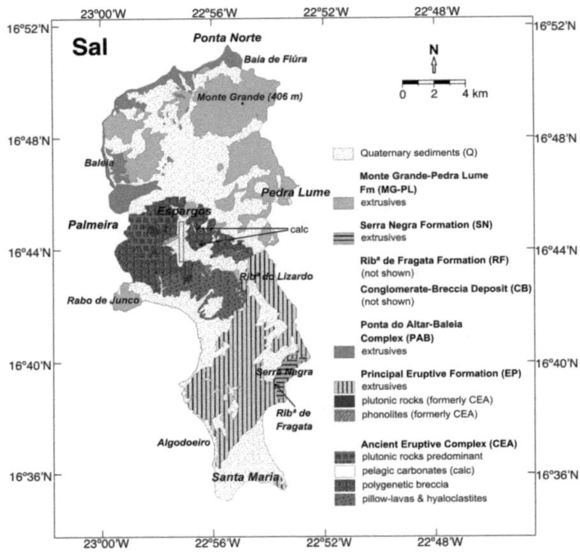

Fig. 5.2 Geological map of Sal, modified from Silva et al. [24] and Holm et al. [21]

and basaltic nature occur in association with the previous lithologies [24], also exhibiting a typical hydroclastic texture (often with glassy matrix) and supporting a submarine environment for the origin of the Ancient Eruptive Complex [19].

In their study [19] state that the final effusions of the Ancient Eruptive Complex correspond to 'small and scarce' fissural subaerial flows with plugs, dykes and flows of phonolites. These outcrops, though, probably belong to the younger Principal Eruptive Formation, due to their apparent discordance with the submarine volcanic products attributed to the Ancient Eruptive Complex. The complexity and weathering of the Ancient Eruptive Complex unfortunately prevent conclusive event reconstructions based on field observations.

Silva et al. [24] noted the presence of small outcrops of dolomitic limestone apparently within the ancient submarine lavas, near Espargos. These sediments occur in a discontinuous outcrop extending over 1,600 m running N–S to NNW–SSE, exhibiting a maximum thickness of about 1.5 m and dipping between 30° and 45° to ENE [19, 32]. The outcrops are presently found at elevations ranging from 40 to 50 m above sea-level (asl). The limestones are micritic mudstones with abundant microfossils. The attribution of an age by biostratigraphy is not simple. A previous study by Ubaldo et al. [32] revealed a planktonic fauna attributed to the Mid Miocene, with reworked Oligocene-Miocene material. According with these authors, faunal preservation is poor and is dominated by globigerinoides, with minor quantities of globigerinas and globoquadrinas. The presence of *Orbulina universa* is mentioned by Ubaldo et al. [32] which would place a lower age constraint of 14.7 Ma on these sediments [33]. Our own study revealed a faunal association dominated by unkeeled planktic foraminifera where *O. universa* seems to be absent. This association would suggest instead a Early Miocene age for these sediments (21.5–16.4 Ma). In agreement, the lack of keeled forms and the absence of *Fohsella spp.*—so abundant in tropical and subtropical Mid Miocene associations—seems to be another indicator of Early Miocene age. However, a single specimen of *Praeorbulina sicana* was found in one of the thin sections made for this study. This species is restricted to Zones N8–N9 (lower part) and would restrict the age of this material to a Langhian age (14.5–17 Ma) [34]. Since the attribution of an age based in a single occurrence is not solid enough, we therefore use a wider age interval between 21.5 and 14.5 Ma, but the possibility for this material to be Langhian with reworked Early Miocene faunas should not be discarded. The sediments—a typical globigerina ooze—attest to the submarine origin of the sequence and provide some constraints regarding the depositional environment of the sequence. In Ubaldo et al. [32] study, the authors did not consider the palaeo-depth at which these carbonates were deposited, or what the depositional process was. It is, however, possible to infer a palaeo-depth using the faunal association present in the sediments. The clear dominance of planktic forms and the relative lack of benthic forms (including larger benthic foraminifera i.e. reef dwellers) suggest depositional conditions deeper than a shelf environment. Furthermore, the high planktic/benthic ratio indicate pelagic conditions associated with depths in excess of 300–400 m [35]. Thus, we support the hypothesis for the Ancient Eruptive Complex to correspond to a seamount sequence and we suggest the sea-level height contemporaneous of the sediments to be in excess of 350 m asl. However, the spatial coexistence and the age overlap between pelagic limestones, apparently interbedded within the Ancient Eruptive Complex sequence, with submarine and subaerial lava flows from the Principal Eruptive Eruption poses a geological problem that is still unresolved.

5.3.1.2 The Principal Eruptive Formation and the Ponta do Altar-Baleia Complex

The Principal Eruptive Formation lies unconformably over the Ancient Eruptive Complex and comprises mainly submarine and subaerial lavas (lower series) and subaerial lavas (upper series), possibly extruded around 16 Ma (lower series) and 9–10 Ma (upper series) [19]. Holm et al. [21] proposed an age range of 16–14 Ma for the lower series, considering the upper series to be part of the Ponta do Altar-Baleia Complex. However, it is more likely that the Principal Eruptive Formation and the Ponta do Altar-Baleia Complex were extruded during the same event, in a wider period between 16 and 9 Ma. This possibility is suggested by the island's volcanic structure and the apparently synchronous available geochronological data [19, 21], a possibility already admitted by Torres et al. [19]. Thus, The Principal Eruptive Formation and the Ponta do Altar-Baleia Complex seem to be the result of the main shield-building stage of the island's evolutionary history.

The Principal Eruptive Formation submarine flows can be found in the lower part of the sequence, probably constituting the oldest exposed rocks of this unit. They can be found exclusively in a small area in the central-eastern part of the island (between Ribeira do Lizardo and Morro da Terra Branca). In this area the submarine sequence consists of a pile of pillow and lobate lavas, embedded in hyaloclastites, all very weathered. It is possible in places to observe the passage zone between subaerial and submarine morphologies, suggesting a contemporaneous water level at $\sim$50 m asl. The precise age of these lavas is still unknown because the geochronological data published so far [19, 21] was determined in the subaerial lavas found several kilometers to the south. The fact that the upper sequence of the Principal Eruptive Formation is presently found as low as the modern sea-level suggests that the relative sea-level height during the age range of its products was below the present level.

The Ponta do Altar-Baleia Complex, comprises from the base to the top: subaerial and submarine flows; shallow marine sediments; subaerial flows (and its vents) [19]. According to Torres et al. [19] this unit extruded in the north and northwest part of the island, between Baía de Fiúra and Palmeira. The basal subaerial lavas presently crop out from 0 to 27 m asl and are covered by submarine flows. These were once extensive (but not very thick) and have been eroded to the point that only isolated outcrops can be found over the basal subaerial flow [19]. These isolated outcrops can be found up to 2 km inland, up to 32 m asl. Thin (normally <1 m) shallow marine carbonates can be found above the previous units (and below the later subaerial flows), exposed only on the coast. They are probably contemporaneous with the erosional period that dismantled the submarine flows. The sediments comprise mainly fossiliferous calcarenites, with lenses of conglomerates, sands and silts [19]. Their fossils include small corals, gastropods, bivalves and echinoid spicules, compatible with the very shallow depths typical of beach deposits. Later thin subaerial flows partially covered the sequence, sometimes surrounding the higher outcrops of submarine lavas [19]. The Ponta do Altar-Baleia sequence suggests contemporaneous small scale variations in the relative sea-level height, at least between 0 and 32 m asl.

5.3.1.3 Intercalated Sediments

An important period of volcanic quiescence and erosion followed the extrusion of the main shield-building stage, as attested to by the erosional contact with subsequent units and the presence of terrestrial and marine sediments—the Conglomerate-Breccia Deposit and the Ribeira de Fragata Formation. The first comprises mud and debris flow deposits and the latter is formed of fossiliferous shallow marine carbonates [19].

The Fragata Formation lies unconformably over the Principal Eruptive Formation subaerial flows, and is covered by the submarine lavas of the Serra Negra Formation [19]. This carbonate constitutes a very regular horizon not exceeding 2 m of thickness, and was deposited in a contemporaneous wave-cut surface presently at $\sim$50 m asl (see Fig. 5.3). These sediments exhibit a conglomeratic basal layer overlain by a compact calcarenite bed. The conglomerate is poorly sorted, graded, with large sub-rounded to well-rounded terrigenous clasts supported by a calcarenitic matrix. Its fossil association includes echinoids, gastropods, bivalves and rare isolated corals. The uppermost layer is a 50 cm thick graded compact calcarenite, exhibiting a grainstone/floatstone texture, and rich in rhodoliths, gastropods and bivalves. These features suggest that the sediments were originally formed in a high energy environment of very shallow depth, but the environment changed to calmer and deeper waters (as suggested by the presence of well articulated rhodoliths and the increase of finer material). The observed transition might be the result of a small relative sea-level rise compatible with the culmination of a highstand. The conglomeratic sediments found at the base of the deposit are thus interpreted as a transgressive lag deposit. Torres et al. [19] attributed an Late Miocene age to these deposits, based on the fossil association, and the age of the overlying submarine volcanic sequence.

Fig. 5.3 Photographs of examples of palaeo-markers of sea-level in the Cape Verdes. **a** Sequence in Ribeira de Fragata valley, in Sal island, comprising from the base to the top: lavas of the Principal Eruptive Fm eroded by a Miocene wave-cut platform and materialised by the marine sediments of Ribeira de Fragata Fm (RF); overlying Serra Negra Fm pillow-lavas and hyaloclastites (SNsub), and Serra Negra Fm subaerial lavas (SN) (with a clear passage zone at ~ 90 m asl); Quaternary terrace at ~ 60 m asl. **b** Echinoid holes in Ponta do Atum, Santo Antão island, at 10–12 m asl. **c** Pillow-breccia and hyaloclastites of Flamengos Fm at ~ 450 m, in Portal da Furna, Santiago Island. **d** Sequence in the northern peninsula of Santiago (Ponta Moreia-Achada Costa region), comprising from the base to the top: Ancient Eruptive Complex (CA), epiclastic carbonatite sediments (CB), calcarenites (calc), Pico da Antónia Volcanic Complex lower member submarine flow(s) at ~ 180 m asl

5.3.1.4 The Serra Negra Post-erosional Volcanic Sequence

The rejuvenation of the volcanic activity briefly took place around 5.5 Ma, creating the Serra Negra Formation [19, 21]. The Serra Negra Formation corresponds to an eroded lava delta whose source vents are no longer distinguishable. The sequence comprise NW dipping foresets of hyaloclastite breccia, with clusters of pillow-lavas, covered by flat-lying subaerial lavas not exceeding a total thickness of 10–15 m. The passage zone is clearly visible and regularly crop out at $\sim 90\,\text{m}$ asl, constituting a solid indication of contemporaneous sea-level height. The subaerial lavas seem to have been extruded around 5.5 Ma [19], so we assume the submarine lavas to be contemporaneous. The Serra Negra Formation lies conformably over the Ribeira de Fragata Formation (see Fig. 5.3).

5.3.1.5 Recent Sediments and Last Post-erosional Volcanic Activity

An erosive period subsequently followed, only interrupted by the extrusion of the latest post-erosional volcanic period between 1 and 0.6 Ma [19, 21]. This volcanic activity originated the Monte Grande-Pedra Lume Formation [19, 21, 24], which comprises a series of isolated strombolian-hawaiian cones with exclusively subaerial products, and a phreato-magmatic structure at Pedra Lume. The products of this stage are only found in the central/northern part of the island and crop out down to the present sea-level.

On the west bank of Ribeira da Fragata, near its source, a marine conglomerate crops out at 70 m asl, above the hyaloclastites of Serra Negra Formation and below fossil sand dunes. These sediments were considered Quaternary by Silva et al. [24] but later Zazo et al. [36] suggested a broader Plio-Pleistocene age. Its presence suggests a probable contemporaneous sea-level height close to 75–80 m asl.

The denuded and low-lying morphology of the island is mostly due to Quaternary marine erosion, as evidenced by the existence of preserved extensive wave-cut platforms and terraces [19, 24, 37–39]. These terraces a distinctive features of Sal's landscape because they cover large areas of the island with a remarkably flat morphology.The terraces (locally called *lajedos*) correspond to beach levels from 2 m up to 100 m asl [19]. Recently Zazo et al. [36] identified 16 different terrace levels from 1 to 60 m asl. Zazo et al. [36] do not mention the 100 m level, on top of Serra Negra, previously identified by Silva et al. [24]. Our study reveals these outcrops to be <30 cm thick compact limestones, weathered, partially recrystallised, and very poor in macro-fossils. In thin section these rocks show an oosparite boundstone texture with poorly sorted, graded, well-rounded grains with several generations of overgrowth of the grains, resulting in typical oncoidal structures— features indicative of a high energy shallow marine environment. The nature of the sediments, together with the regular flat-lying morphology of the top surface of Serra Negra where the sediments lye, suggests the existence of a wave-cut surface and terrace in this place. Thus, in our view, the 100 m terrace reflects a contemporaneous sea-level position (the highest Plio-Pleistocene terrace in Sal) and should be included in freeboard reconstructions. We propose that this terrace is older than 1 Ma because this is the oldest age obtained for the Monte Grande-Pedra Lume subaerial lavas, found topographically lower than the terrace and with no apparent signs of significant marine erosion. In fact, the Monte Gordo-Pedra Lume products are only covered by the <25 m terraces.

5.3.2 *Boa Vista*

One of the best examples of a razed volcano within the archipelago, Boa Vista is surprisingly flat, practically devoid of young volcanic forms [40]. Its morphology is only interrupted by a few residual prominences, mostly eroded phonolitic edifices, whose tallest barely reaches 380 m asl. The erosion experienced by this island allows a unique observational window into the internal structure of a volcanic edifice. Boa Vista is also the Cape Verdean island with the highest felsic/mafic volumetric ratio [40].

Absolute ages are unknown, but the island's volcanostratigraphy was defined by Serralheiro et al. [40] and comprise, from the base to the top: Ancient Eruptive Complex; Monte Passarão Complex; Fundo de Figueiras Formation; Monte Caçador Formation; Pico Forcado Formation; Intercalated Sedimentary Deposits; Chão de Calheta Formation; Modern Pyroclastic Cones; Pleistocene Sediments; and Holocene Sediments (see Fig. 5.4, and Table 5.2 at the end of this chapter).

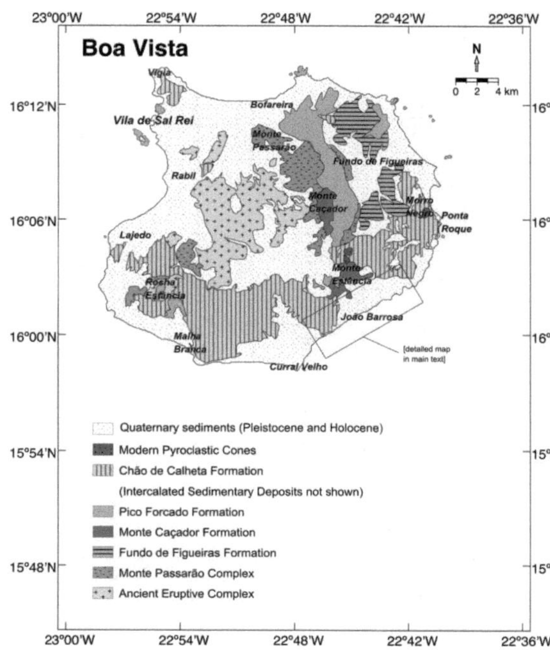

Fig. 5.4 Geological map of Boa Vista, after Serralheiro et al. [40]

Table 5.2 Summarized volcanostrigraphy of Boa Vista, with age and sea-level constraints

Units[a,b]	Main characteristics	Age data for exposed rocks	Evidence for contemporaneous sea-level	Palaeo-depth constraints	Sea-level constraints	Key localities for sea-level markers
Holocene sediments	Beach sand, alluvium, colluvium, etc	Modern deposits	–	–	–	–
Erosional contacts						
Pleistocene sediments	Extensive marine terraces (and wave-cut surfaces) and contemporaneous dunes	considered as *Pleistocene*[a]	Marine terraces and wave-cut surfaces ranging 1–105 m asl (or even 130 m asl)	Beach deposits and shallow marine sediments; inferred palaeo-depth <5 m	Sea-level up to 110 m asl	Malha Branca; Curral Velho; João Barrosa; Rabil-Lajedos; Bofareira; Vigia
Erosional contact						
Modern Pyroclastic Cones	Post-erosional stage strombolian cones and associated flows[a]	Pleistocene?[a]	Unit is subaerial and crops out above 50 m asl	–	Sea-level below 50 m asl	–
Erosional contact						
Chão de Calheta Fm	Effuside submarine and subaerial basalts[a]	Pliocene?[a]	Submarine lavas up to 30–40 m asl	Passage zone sometimes observable at 30–40 m asl	Sea-level up to 30–40 m asl	Eastern coast near Ponta Roque and region inland
Conformable contact						
Intercalated Sedimentary deposits	Mostly shallow marine sediments; volcanic quiescence and erosion[a]	Pliocene?[a]	Marine sediments cropping out up to 10 m asl	5–30 m	Sea-level up to 30–40 m asl	Eastern coast near Ponta Roque and region inland

(continued)

Table 5.2 (continued)

Units[a,b]	Main characteristics	Age data for exposed rocks	Evidence for contemporaneous sea-level	Palaeo-depth constraints	Sea-level constraints	Key localities for sea-level markers
Erosional contact						
Pico Forcado Fm	Subaerial phonolitic flows and pyroclastic products[a]	Unknown	Unit possibly entirely subaerial, cropping down to present sea-level	–	Sea-level below present sea-level	–
Erosional contact						
Monte Caçador Fm	Effusive phonolitic series, probably subaeria[la]	Unknown	Unit possibly entirely subaerial, cropping down to present sea-level	–	Sea-level below present sea-level	–
Erosional contact						
Fundo de Figueiras Fm	Comprise effusive submarine and subaerial basalts[a]	Unknown	Submarine flows crop out to ~35 m asl	Unknown	Sea-level up to ~35 m asl	Malha Branca, Fundo de Figueiras
Nature of contact not clear						
Monte Passarão Complex	Phonolitic series of subaerial flows, dykes, breccia and ignimbrites. Maybe is part of Ancient Eruptive Complex[a]	Unknown	Unit possibly entirely subaerial, cropping down to ~50 m asl	–	Sea-level below ~50 m asl	Unknown
Nature of contact not clear						
Ancient Eruptive Complex	Extrusive phonolitic rocks, intensely intruded by large plutons of nepheline syenites[a]	Unknown	Unknown	–	Unknown	Unknown

[a] Serralheiro et al. [40]
[b] Serralheiro 1976 [22]

The island exposed geology seemingly documents a dominantly subaerial evolution since the majority of the submarine markers do not exceed 40 m asl. However, the edifice was subjected to extensive submersion during the Quaternary, judging by the presence of preserved marine erosive features found. As a result, Boa Vista is one of the islands of the archipelago whose surface is more extensively covered with sedimentary rocks (mostly Holocene sands and Pleistocene terraces).

5.3.2.1 The Ancient Eruptive Complex

The Ancient Eruptive Complex of Boa Vista occupies a large portion of the island's central surface. This unit is characterised by rocks of phonolitic nature, intensively intruded by dykes and large pockets of nepheline syenites with exotic minerals [40]. Sea-level constraints are unknown.

5.3.2.2 The Monte Passarão Complex

The following unit corresponds to the Monte Passarão Complex, a phonolitic series of subaerial flows, dykes, breccia and ignimbrites. This unit crops out around the margins of the Ancient Complex in the central area of the island, although the geometric relation with the latter unit is unclear [40]. Unit is possibly entirely subaerial and its products occur down to $\sim$50 m asl, so contemporaneous sea-level was below this elevation.

5.3.2.3 The Fundo de Figueiras Formation

The Fundo de Figueiras Formation comprises mostly effusive basaltic, submarine and subaerial products, and a dyke swarm [40]. The submarine flows crop out mostly between Rocha Estância and Malha Branca, up to $\sim$35 m asl. They are rare in the rest of the island and their stratigraphic relation to the subaerial flows is not completely understood. Despite its small outcrop it is possible to infer that the relative sea-level height when these submarine flows were erupted was at minimum 35 m asl.

5.3.2.4 The Monte Caçador Formation

The Monte Caçador Formation, a generally effusive phonolitic series, is unconformably over the Monte Passarão phonolites and the Fundo de Figueira basalts Serralheiro et al. [40]. This unit seems to be entirely subaerial, exposed down to elevations close to the present sea-level.

5.3.2.5 The Pico Forcado Formation

The Pico Forcado Formation is yet another phonolitic unit, mostly subaerial flows and a few pyroclastic products [40]. The majority of the relief on the island, especially the eastern summits, are isolated structures and products belonging to this unit. It crops out in various positions, some close to the present sea-level.

5.3.2.6 Intercalated Sediments

A period of volcanic quiescence and erosion is suggested by the presence of terrestrial and marine sediments above the previous units and below the following ones. The marine deposits are found mostly on the coast, above the Pico Forcado and below the submarine lavas of the Chão de Calheta Formation, and can be followed further inland, through the more incised valleys [40]. They crop out close to the modern sea-level (<10 m asl). These rocks are calcarenites and compact limestones, fossiliferous, sometimes reach 10 m of thickness, and correspond to shallow water deposits [40].

5.3.2.7 The Chão de Calheta Formation

According to Serralheiro et al. [40] the next unit—the Chão de Calheta Formation—is the the most important basaltic unit found on the island. This effusive series, subaerial and submarine, is only found on the periphery of the island where it covers large areas. The submarine flows are mostly found on the coast (particularly in the east), frequently underlain by the marine sediments, and reach considerable individual thicknesses [40]. They crop out up to 30–40 m asl, and the passage zone to the overlying subaerial flows is frequently observable.

5.3.2.8 Recent Post-erosional Volcanic Activity

A few rare modern pyroclastic cones of basaltic nature can be found in the island (e.g. Morro Negro), with small associated flows [40]. These cones are only found inland above 40–50 m asl and submarine products belonging to this stage seem to be absent.

5.3.2.9 The Pleistocene Sediments and Wave-cut Surfaces

Above the previous units, and covering extremely large areas, the Pleistocene marine terraces can be found up to 8 km inland and up to elevations in excess of 100 m asl. Serralheiro et al. [40] describe terraces up to 130 m asl. Our study revealed a series of different levels (although some may correspond to the stacking/overlapping of several individual terraces): 2–3, 4–8, 10–15, 18–28, 30–45, 50–60, 65–70, 70–80, and 90–105m asl (and eventually 130 m asl). These elevation intervals can occasionally be more extensive, and several generations of Pleistocene dunes can also be found on top of some of these levels, making it difficult (due to recrystallisation and calcrete formation) to pinpoint the top of the marine sediments. The terraces vary from 50 cm up to 15 m of thickness and are normally characterised by a gentle seaward dipping top surface (between 1° and 3°) interpreted as a geomorphological remnant of the shoreface. Cliffed edges sometimes truncate the deposit, as a result of later subaerial (headward erosion and downwearing of slopes) but mostly due to marine erosion (regressive wave-cut truncations). In contrast with this usual morphology, in some places it is possible to find steeply dipping deposits, normally of smaller extent and thickness. These are composed of grainstones exhibiting plane-bedding and a top surface dipping 20°–25° seaward (e.g. near Ponta do Medronho, at 25–45 m asl). Their occurrence along constant elevations and the presence of trace-fossils (vertical burrows of the ichnogenus *Ophiomorpha* or *Thalassinoides*) suggest these sediments to be marine rather than aeolian in origin. We interpret these as steep foreshore beach deposits.

The great majority of the Pliocene terraces are composed of well-cemented bioclastic grainstones, wackestones, boundstones, and mudstones, frequently re-crystallised and variably affected by processes of calcrete formation (typical of this type of arid environment). Internal sedimentary structures are sometimes observable, normally dominated by plane-bedding and lamination and less frequently by cross-bedding and cross-lamination. The construction of one deposit at the expense of previous (normally higher) deposits, by erosion and redeposition, probably occurred in places. The evidence is the presence of a conglomeratic facies rich in lithoclasts/intraclasts at the base of the foreshore face built in the older deposits. Conglomeratic and micro-conglomeratic facies comprising well rounded volcanic clasts and occasional lithoclasts/intraclasts supported by a calcarenitic matrix can also be found generally at the base of some terraces or at the base of steep foreshores built in the igneous bedrock, indicating higher energy environments. Fossil associations include solitary corals (in the high energy facies), rhodoliths, bivalves, gastropods, and occasional echinoderms. Trace-fossils (burrows identified as belonging to the ichnogenus *Palaeophycus*) are also very common, especially in the wackstone/mudstones facies, normally occurring atop the more extensive deposits (e.g 10–15 m level). Good areas to observe the terraces include: all the south coast, the Rabil-Lajedos, Bofareira and Vigia. The Rabil-Lajedos ridge (40–100 m asl) is notable for the extent and thickness of the carbonate sediments that constitute its crest, due to stacking of several generations of terraces and dunes.

In the south coast, from Malha Branca (Santa Mónica beach) to Curral Velho and João Barrosa, the terraces feature an outstandingly clear stair-step morphology (similar to the deposits of Sal and Maio), allowing the reconstruction of the coastal evolutionary history during the Quaternary (see Fig. 5.5).

The overall features suggest the terraces may be interpreted as open-shelf to wave-dominated beach deposits, formed by processes similar to those occurring in present coastal areas. Thus, these levels constitute good tracers of relative past sea-

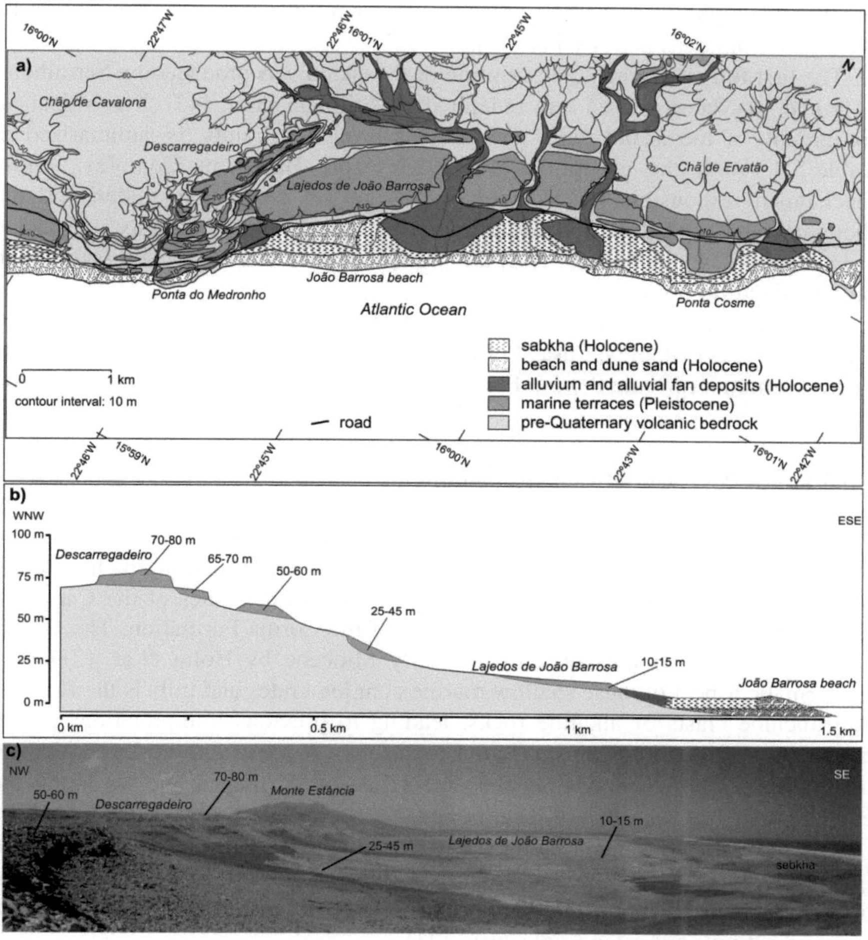

Fig. 5.5 **a** Simplified geological map of João Barrosa area, in SE Boa Vista, showing the Quaternary marine terraces; **b** Schematic cross-section from Descarregadeiro to João Barrosa beach, showing the position of the Quaternary marine terraces (key as the map); **c** Photo of the same area

level and the elevations at which these deposits occur suggest several positions of Quaternary relative sea-level heights of up to 100–110 m asl.

5.3.3 Maio

Completely devoid of young volcanic morphologies, Maio is a small and low-lying island where central elevations only reach 430 m asl. The island is famous for the exposure of Mesozoic fossiliferous marine sediments [41] and MORB basalts [42] regarded to be part of a large fragment of ocean floor brought up by the island's igneous activity [41, 43].

The first full stratigraphic description of the island was produced by Serralheiro [41] and was later revised and redefined by Stillman et al. [43]. The volcanostratigraphy of the island, with age and sea-level constraints, is summarized in Table 5.3. With the exception of the Mesozoic Basement Complex, whose mechanism responsible for its upheaval is not yet completely understood but thought to be related to updoming associated with plutonic activity and island volcanism [41, 43], the majority of the pre-Quaternary markers of sea-level found on the island do not exceed 40 m asl.

According to Stillman et al. [43], it includes: Mesozoic Basement Complex; Central Igneous Complex; Casas Velhas Formation; Pedro Vaz Formation; Malhada Pedra Formation; Monte Penoso Formation; Quaternary Sediments (see Fig. 5.6).

5.3.3.1 The Mesozoic Basement Complex

The Mesozoic Basement Complex occupies up to 20% of the island's area and includes: the MORB type volcanic rocks of the Batalha Formation, the limestones of the Morro Formation, the shales and thin-bedded limestones of the Carqueijo Formation, and the tuffs and conglomerates of the Coruja Formation. The Coruja Formation was later attributed to the Early Miocene by Holm et al. [21]. The problematic aspect of these shallow marine conglomerates and tuffs is the fact that they include clasts of alkaline rocks, leading to previous interpretations for an earlier start of island volcanism (Late Cretaceous). A constraint on the correct age of this unit is crucial to dating the emergence of the island and to dating the contemporaneous relative sea-level height.

The Mesozoic Basement Complex is folded and faulted, flanking the Central Igneous Complex. The mechanism responsible for its upheaval is not yet completely understood, but is thought to be related to updoming associated with plutonic activity and island volcanism [41].

5.3.3.2 The Central Igneous Complex

The Central Igneous Complex corresponds to a series of intrusive bodies composed of pyroxenites, essexites and syenites, densely cut by a swarm of sills and dykes of varied composition [41, 43]. The age of these intrusions is still debatable but the 18–21 Ma interval seems to best represent it [21].

5.3.3.3 The Casas Velhas Formation

The Casas Velhas Formation seems to correspond to the first extrusive stage of Neogene volcanism [43]. This unit mostly crops out in the valleys of the SW of the

Table 5.3 Summarized volcanostrigraphy of Maio, with age and sea-level constraints

Units[a]	Main characteristics	Age data for exposed rocks	Evidence for contemporaneous sea-level	Palaeo-depth constraints	Sea-level constraints	Key localities for sea-level markers
Quaternary sediments	Extensive marine terraces (and wave-cut surfaces), dunes, alluvium, colluvium, etc	*Quaternary*[b]	Marine terraces and wave-cut surfaces ranging at 2–6, 8–12, 15–20, 30–40, 55–65, 65–75, and 80–100 m asl	Beach deposits and shallow marine sediments. Inferred palaeo-depth <5 m	Sea-level up to 105 m asl	Riberia do Morro; Praia Gonçalo-Santo António region
Erosional contact						
Mio-Pliocene Marine Sediments[b]	Shallow marine sediments of small expression	Inferred as Mio-Pliocene due to fossil content[b]	Shallow marine sediments up to 1 m asl	Shallow marine sediments. Inferred palaeo-depth 5–10 m	Sea-level up to 5–10 m asl	Vila do Maio
Erosional contact						
Malhada Pedra/ Monte Penoso Fm	Extensive subaerial effusive products, and the remains of a strato-volcano[a,c]	Between 7.3 ± 0.4 and 6.5 ± 1.2 Ma (K/Ar)[c]	Unit possibly entirely subaerial cropping out down to present sea-level	–	Sea-level below present level	–
Conformable contact?						
Pedro Vaz Fm	Mostly fanglomerate sdiments and possible marine sediments with possible subaerial lavas interbedded[a]	–	Alternation between marine and fluvial sediments with possible subaerial lavas[a] up to 50 m asl	Alternation between subaerial and submarine facies. Proximity to sea-level	Sea-level oscillations around 0–50 m asl	Pedro Vaz

(continued)

Table 5.3 (continued)

Units[a]	Main characteristics	Age data for exposed rocks	Evidence for contemporaneous sea-level	Palaeo-depth constraints	Sea-level constraints	Key localities for sea-level markers
Erosional contact						
Casas Velhas Fm	First extrusive stage of Neogene volcanism. Effusive submarine and subaerial flows extruded in a littoral environment[a]	Extruded around 9.8 ± 0.8 Ma (K/Ar)[c]	Submarine morphologies found up to 35–40 asl	Possible passage zone around 35–40 m asl	Sea-level up to 35–40 m asl	Ribeira de Cumiassa; Ribeira das Casas Velhas
Conformable, intrusive or tectonic, depending on localities						
Central Igneous Complex	Series of intrusive bodies of essexites and syenites, intensely intruded by a dyke swarm[a,b]	Possibly between 18 and 21 Ma[d] based on K/Ar[c]	–	–	–	–
Intrusive or tectonic contact						
Mesozoic Basement Complex	MORB-type volcanic rocks and Jurassic/Cretaceous sediments. Interpreted as uplifted seafloor[a,b]	Upper Jurassic to Early Upper Cretaceous Biostratigraphy)[a,b,c]	Pelagic sediments and MORB basalts	Pelagic faunas	Sea-level in excess of 4,000 m asl	Ribeira do Morro; Ribeira de D. João

[a] Stillman et al. [43]
[b] Serralheiro [41]
[c] Mitchell et al. [20]
[d] Holm et al. [21]

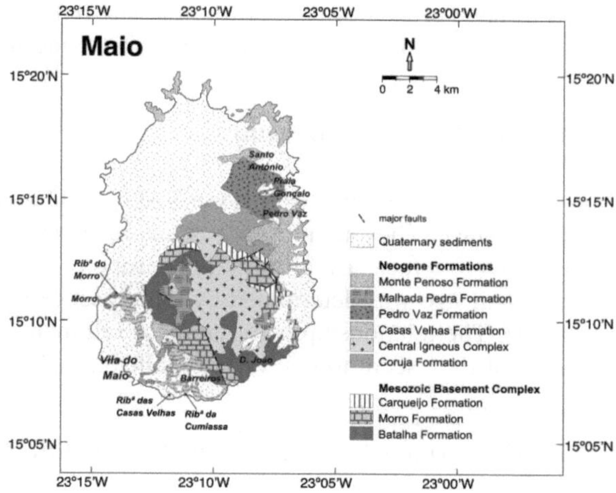

Fig. 5.6 Geological map of Maio, after [41, 43] and Holm et al. [21]

island, and comprises submarine and subaerial lavas, according to Stillman et al. [43]. These authors state that the proximity of a Neogene coastline is suggested by the presence of lava delta deposits with hyaloclastites and isolated pillows, with rare subaerial flows atop. Since these submarine flows crop out at elevations up to 35–40 m asl we infer an elevation of ~ 40 m for the sea-level position. This unit was apparently emplaced sometime around 10 Ma [20].

5.3.3.4 The Pedro Vaz Formation

The Pedro Vaz Formation comprise conglomerates and tuffs synchronous of the Casas Velhas Formation, representing a series of immature fluvial fanglomerates with possible marine sediments [43]. The coexistence of subaerial and submarine facies can be explained by rapid vertical movements [43], or more simply by the high-frequency variations that characterise the eustatic curve.

5.3.3.5 The Malhada Pedra Formation

The Malhada Pedra Formation is the result of extensive subaerial volcanism, creating large quantities of plateau lavas around 9–7 Ma, unconformably overlying the previous units [20, 43]. The products of this volcanic stage can presently be found down to lower elevations, close to the present sea-level.

5.3.3.6 The Monte Penoso Formation

Following the effusive volcanism of Malhada Pedra, the Monte Penoso Formation corresponds to the remains of a prominent strato-volcano, now eroded [43]. This seems to have erupted around 7 Ma, the last reported age for volcanic activity in Maio [20]. These products can presently be found down to present sea-level.

5.3.3.7 Mio-Pliocene Sediments

Mio-Pliocene calcarenites were described by [41] on the coast to the south of Vila do Maio village. They are presently at sea-level and are unconformably covered by a Quaternary terrace. Bivalves and gastropods fossils are present in these sediments. Their age was determined by the presence of *Procarcharodon megalodon* teeth [41]. The relationship between these sediments and the volcanic units is still unclear but they seem to overly the Malhada Pedra lavas.

5.3.3.8 Quaternary Sediments and Wave-cut Surfaces

Like Boa Vista and Sal, the island of Maio is also covered by extensive Pleistocene and Holocene deposits. The Pleistocene deposits are shallow marine terraces with elevations ranging from 2 to 100 m, according to Serralheiro [41]. This author described them as "discontinuous rings around the centre of the island, in a staircase arrangement, with a regular flat surface slightly inclined to seaward". The same author identified 6 main levels at 2–6, 8–12, 15–20, 30–40, 50–60, and 80–100 m, with associated contemporaneous dune systems. Some of these levels correspond, though, to the stacking of several terraces and it is frequently possible to observe sedimentological evidence for the construction of one terrace at the expense of an older, slightly higher, terrace. Our field study suggests the existence of seven main terrace levels (instead of 6), at 2–6, 8–12, 15–20, 30–40, 55–65, 65–75 and 80–100 m. The terraces mostly comprise well-cemented bioclastic calcarenites and micro-conglomerates, interpreted as open-shelf to wave-dominated beach deposits. Shallow marine limestones, rich in in situ corals grown on the subjacent lavas, large bivalves and gastropods are also frequent and correspond to barrier/reef environments.

5.3.4 Santiago

Santiago is currently the largest island in the archipelago, and is considered to be an elongated shield-volcano in an early post-erosional stage (Miocene?-Quaternary) [21]. The island reaches its highest point at Pico da Antónia (1394 m) and, despite being cut by deep valleys, its topography features a young morphology where the shape and dimensions of the shield edifice are still perceptible.

The volcanostratigraphy of the island comprises the following main units, from the oldest to the youngest [22]: Ancient Eruptive Complex; Flamengos Formation; Orgãos Formation; Pico da Antónia Eruptive Complex; Assomada Formation; Monte das Vacas Formation; and Quaternary sediments (see Figs. 5.7, 5.8). Recent

Fig. 5.7 Geological map of Santiago, after Serralheiro [22]

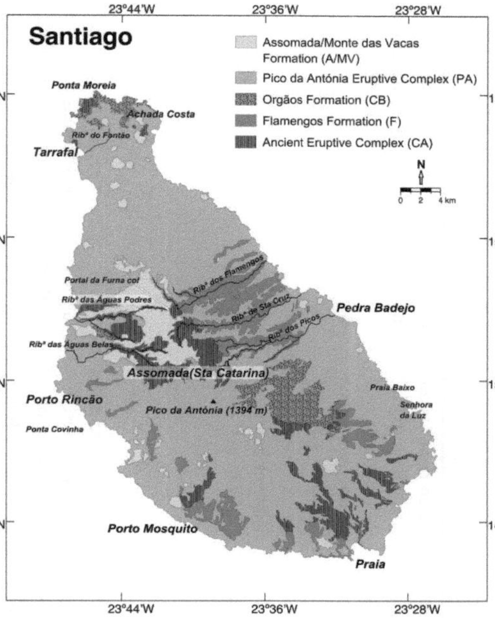

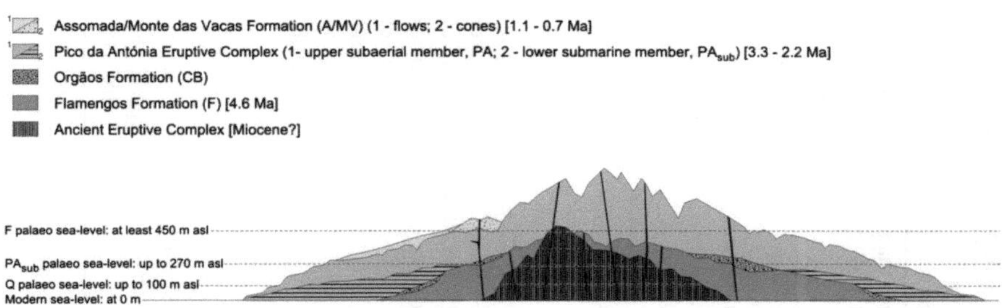

Fig. 5.8 Idealized cross section of Santiago showing the geometric relations between the units, and the information of relative past sea-levels inferred from the volcanostratigraphy. Note the increase in elevation with increasing age. Stratigraphic units after Serralheiro [22], and the geochronology after Holm et al. [21]

radiometric studies suggest, however, that Assomada and Monte das Vacas formations are likely to be different facies of the same volcanic episode [21]. The volcanostratigraphy of the island is summarised in Table 5.4, with age and sea-level constraints.

5.3.4.1 The Ancient Eruptive Complex

The Ancient Eruprive Complex of Santiago constitutes the basement unit of the island, and comprise several distinct products of basaltic, phonolitic and carbonatitic

Table 5.4 Summarized volcanostrigraphy of Santiago, with age and sea-level constraints

Units[a]	Main characteristics	Age data for exposed rocks	Evidence for contemporaneous sea-level	Palaeo-depth constraints	Sea-level constraints	Key localities for sea-level markers
Quaternary sediments	Marine terraces (and wave-cut surfaces), alluvium, colluvium, etc	Younger than 1.8 Ma?	Marine terraces and wave-cut surfaces ranging at 2–4, 5–10, 15–25, 30–40, 60–80, and 80–100 m asl	Beach deposits and shallow marine sediments. Inferred palaeo-depth <5 m	Sea-level up to 105 m asl	Porto Rincão, Porto Mosquito, Praia Baixo, Senhora da Luz
Erosional and conformable contact, depending on localities						
Assomada/ Monte das Vacas Fm	Post-erosional stage composed of strombolian-hawaiian cones and valley-filling efusive sequences	Between 1.13 ± 0.03 and 0.74 ± 0.04 Ma (Ar/Ar)[c]	Entirely subaerial, but only present above 40 m asl	Sea-evel below 40 m asl	Sea-level below 40 m asl	–
Erosional contact						
Pico da Antónia Eruptive Complex	Main shield-building stage, mostly comprising effusive products. Possibly composed of two distinct members[c]: a lower member, entirely submarine; an uppe subaerial member with submarine morphologies in the fringes of the edifice	Between 3.25 ± 0.04 and 2.25 ± 0.09 Ma (Ar/Ar)[b]	Lower member entirely submarine and cropping out up to 270 m asl; upper member with submarine morphologies in the fringes of the edifice, generally below 100 m asl	Passage zone is not well exposed. Elevation of highest point assumed as possible maximum sea-level height	Sea-level up to 270 m asl	Lower region of Ribeira dos Flamengos, Ribeira de Santa Cruz, Ribeira dos Picos, Ribeira das Águas Belas; Porto Mosquito; Porto Rincão; Northern peninsula
Erosional and conformable contact, depending on localities						
Órgãos Fm	Terrestrial and marine conglomerate deposit[a]	Unknown	Deposits with marine characteristics	Unknown	Unknown	–

(continued)

Table 5.4 (continued)

Units[a]	Main characteristics	Age data for exposed rocks	Evidence for contemporaneous sea-level	Palaeo-depth constraints	Sea-level constraints	Key localities for sea-level markers
Erosional contact						
Flamengos Fm	Exclusively submarine lavas.[a] Possibly early submarine shield-building stage	Between 4.59 ± 0.09 and 4.57 ± 0.31 Ma (Ar/Ar)[b]	Entire unit is submarine, cropping out up to 450 m asl	Top is eroded. Elevation of highest point assumed as minimum sea-level height	Sea-level in excess of 450 m asl	Highest point in Portal da Furna Col (Assumada); Flamengos Valley
Erosional contact						
Ancient Eruptive Complex	Severely altered volcanic products of possible submarine origin. Intensely intruded by dykes and pockets of plutonic rocks.[a] Possible existence of fragments of MORB-type basalts.[d] End of seamount stage?	Poorly constrained? Late Miocene? 9.8 ± 0.6 and 8.5 ± 1.4 Ma in a mica from carbonatites (K/Ar)[e]	Entire unit is possibly submarine, but complexity prevents solid event reconstructions	Unknown	Unknown	–

[a] Serralheiro [22]
[b] Holm et al. [21]
[c] Martins et al. [46]
[d] Gerlach et al. [44]
[e] Bernard-Griffiths et al. [59]

nature. The whole unit is deeply eroded, weathered, and intensely intruded by a dyke swarm and pockets of plutonic rocks [22]. The Ancient Eruptive Complex country rock is rarely observed but seems to be mostly constituted by submarine volcanic products. Fragments of sea-floor (MORB-type basalts) seem to be present within the sequence of the Ancient Eruptive Complex, as suggested by Gerlach et al. [44] and Davies et al. [45], dragged by a process that is still poorly understood. Hence, the Ancient Eruptive Complex may represent an early submarine edifice-building phase of the island [22] possibly including fragments of uplifted sea-floor [21]. It is out of the scope of this work to reconstruct all the events that took place to form this unit, and to unfold its complexity. Thus, we do not take into account inferences of sea-level based in the Ancient Eruptive Complex of Santiago.

5.3.4.2 The Flamengos Formation

The Flamengos Formation is exclusively composed of submarine lavas (pillow-lavas and hyaloclastites) and seems to correspond to an early submarine shield-building stage. It lies unconformably over the Ancient Complex, and presently crop out from the present sea-level up to an elevation of $\sim$450 m asl [22]. This unit constitutes the largest portion the exposed edifice, apparently rapidly extruded around 4.6 Ma [21] but intensively eroded afterwards [22]. The highest point where the Flamengos Formation crops out is in the Portal da Furna col (Assomada). Here, the headward erosion of the Ribeira das Águas Podres and Flamengos valleys exposes the submarine sequence up to 450 m asl. The present sequence exhibits clusters of individual pillows floating in a weathered hyaloclastitic breccia, unconformably covered by subaerial lavas corresponding to the Pico da Antónia Volcanic Complex and Assomada/Monte das Vacas Formations. Since the younger subaerial flows are filling a palaeo-topography superimposed on the Flamengos Formation sequence, it is not possible to quantify how much of this unit was eroded, thus making it impossible to estimate the palaeo-depth of emplacement for the submarine volcanism. Nevertheless an elevation of $\sim$450 m asl should be considered as the minimum height for the palaeo sea-level. Due to its nature and extent, this unit is the oldest most solid constraint of sea-level height in Santiago.

5.3.4.3 The Orgãos Formation

The Orgãos Formation, also known as the Conglomerate-Breccia, comprises thick, subaerial and submarine conglomerate deposits [22]. The unit marks an important erosional period.

5.3.4.4 The Pico da Antónia Volcanic Complex

The Pico da Antónia Eruptive Complex corresponds to the main shield-building stage, erupted between 3.3 and 2.2 Ma [21]. This unit defines the island general

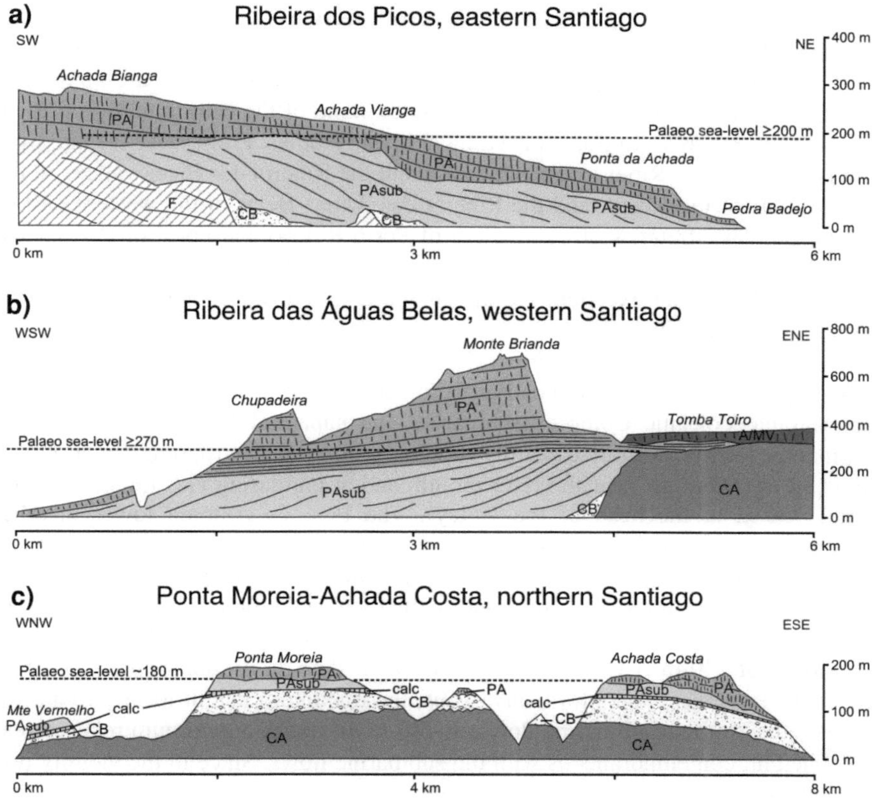

Fig. 5.9 Schematic cross sections of Santiago, showing inferred palaeo sea-levels (key follows): **a** Lower course of Ribeira dos Picos (eastern Santiago); **b** Lower course of Ribeira das Águas Belas (western Santiago); **c** Ponta Moreia—Achada Costa (Tarrafal) region, the northernmost headland of Santiago (adapted from Silva et al. [47] (see Fig. 2 for photographs). **Key**: CA—Ancient Eruptive Complex, CA'—Ancient Eruptive Complex (carbonatitic/nephelinitic extrusive complex), F—Flamengos Formation (submarine lavas), CB—Conglomerate-breccia (marine), CB'—epiclastic carbonatitic sediments (marine), calc—calcarenites, PAsub—submarine flows of Pico da Antónia Volcanic Complex, PA—subaerial flows of Pico da Antónia Volcanic Complex, A/MV—subaerial flows of Assomada/Monte das Vacas Formation

morphology, and mainly comprises voluminous piles of effusive products. The unit seems to be composed of two distinct members generally separated by an erosional contact [46]: (1) a lower member, apparently exclusively submarine; (2) an upper subaerial member with occasional submarine morphologies in the fringes of the edifice and at low elevations. The position of the submarine lower member can particularly be used to track the sea-level height. Its products crop out up to an elevation of ~ 270 m asl, and minor marine carbonate beds are sometimes present at the base or intercalated within the sequence. We therefore infer a minimum palaeo-position of sea-level at ~ 270 m asl, contemporaneous with these submarine flows. In Fig. 5.9 three key cross-sections are shown, marking the position of the palaeo sea-level. A detailed description of key localities follows.

The submarine products of Pico da Antónia Volcanic Complex are ubiquitous along the coast. However, the lower course of some of the central valleys present the best natural cross sections for sea-level height studies. From these we highlight the Flamengos, the Ribeira de Santa Cruz, and the Ribeira dos Picos valleys in the eastern part of the island, and the Ribeira das Águas Belas in the western part of the island. The stretches of coast south of Porto Rincão and north of Porto Mosquito in the western part of the island, are also good examples. The northern tip of the island is also an excellent region for sea-level palaeo-marker studies. These localities are described below.

In the mouth of the Flamengos valley, the exposed sequence reveals a palaeo-topography carved in Flamengos Formation submarine lavas and covered by Pico da Antónia Volcanic Complex submarine and subaerial lavas. The Flamengos Formation locally exhibit seawardly dipping pillow-lavas, cropping out at elevations up to 100 m asl near the river mouth, and at increasing elevations upstream (until 450 m at the head of the valley). The Pico da Antónia lower member submarine lavas are found only at the very end of the valley, "fossilizing" the palaeo-topography imposed on the Flamengos lavas by a very steep contact suggesting a former coastline. In this locality the Pico da Antónia submarine flows are characterised by very steep dipping pillows (to seaward) and can be found up to elevations of 140 m. The later subaerial flows of Pico da Antónia (upper member) cover the sequence, unconformably. Thus, the maximum height locally found for the Pico da Antónia submarine flows, at 140 m asl, mark a minimum relative sea-level height, contemporaneous with the submarine flows. Since in the vicinity it is possible to find the younger subaerial flows at lower elevations, one may infer that sea-level was lower during the emplacement of these flows.

The west bank of Ribeira de Santa Cruz (also called Ribeira da Boaventura) is another key cross section for freeboard studies. The bottom of the valley is carved in Flamengos Formation pillow-lavas and submarine pyroclasts. The top of this unit rises steeply further upstream, in the Ribeirão Boi area, and is covered by a thick sequence of Pico da Antónia lower member submarine flows, again filling the palaeo-topography. Thin marine conglomerates and calcarenites are reported between the two submarine units [22]. The highest outcrops of Pico da Antónia lower member submarine flows can be found near Ribeirão Fundo, at $\sim$235 m asl, or at $\sim$245 m asl in the adjacent valleys of Ribeirão Seco and Carriçal. There, they are steeply against the older submarine flows and are unconformably covered by the later Pico da Antónia subaerial flows (upper member). We thus infer a minimum palaeo-position for contemporaneous sea-level at $\sim$250 m asl.

A similar situation can be found in the Ribeira dos Picos valley, where the Pico da Antónia lower member submarine flows yet again cover a steep palaeo-topography imposed on the Flamengos Formation, and whose top can be found in both banks at $\sim$225 m (Achada Vianga) and $\sim$200 m (João Toro), covered by the later subaerial flows (see Fig. 5.9). This geometry suggests a similar contemporaneous sea-level height above 230 m asl.

The western flank of Santiago island is also fertile ground for freeboard studies. The lower course of the Ribeira das Águas Belas (see Fig. 5.9) is entirely carved in a sequence comprising Pico da Antónia lower member submarine flows unconformably covered by subaerial lavas belonging both to Pico da Antónia Volcanic Complex (upper member) and Assomada/Monte das Vacas Formations. The submarine sequence extends close to 5 km inland from the river mouth, and is constituted of steeply inclined pillow-lavas, pillow breccias, and hyaloclastites, dipping 25°–35° seaward. Its top is eroded but its highest point can be found in Chão do Penedo area, convincingly up to $\sim$270 m, and possibly up to 290 m or even 320 m asl (a slope deposit covers the boundary). We therefore infer a minimum palaeo-position of sea-level at $\sim$270 m asl, contemporaneous with these submarine flows.

The coast between Porto Rincão and Ponta da Covinha exhibits a sequence composed of Pico da Antónia lower member submarine flows unconformably covered by two sequences of subaerial flows (both interpreted to belong to the Pico da Antónia Volcanic Complex upper member). The later subaerial sequence spilled to lower levels, partially covering the boundary between the submarine and first subaerial series. This boundary is irregular but clearly visible in places, indicating erosion prior to the extrusion of the succeeding subaerial flows, marked by the occasional presence of a thin conglomerate bed (<2 m) with large rounded boulders. The top of the submarine flows reaches up to 80–90 m asl, and up to higher elevations further inland along the valleys.

The Porto Mosquito area presents a similar sequence. The exposed sequence comprises a lower package composed of pillow-lavas and thin hyaloclastite beds, dipping seaward and reaching an elevation of $\sim$80 m, and an upper package of subaerial flows (both attributed to the upper member of Pico da Antónia Volcanic Complex). This elevation is interpreted as a minimum height for the contemporaneous relative sea-level.

The northern peninsula of Santiago is a plateau whose sequence comprises, from the base to the top [47]: a carbonatitic/nephelinitic extrusive complex (Ancient Complex); shallow marine epiclastic carbonatitic sediments; bioclastic calcarenites; and Pico da Antónia plateau basalts (see Fig. 5.3). The epiclastic sediments and the bioclastic calcarenites are best observed on the coast between Achada Costa and Ponta Moreia, and in the Ribeira do Fontão valley. The thickness of the sedimentary package is variable but can locally reach >100 m The bioclastic calcarenites are particularly abundant in Ribeira do Fontão valley, where they crop out intercalated with pillow-lavas. These sediments are frequently conglomeratic at the base, with large volcanic boulders and become more calcarenitic toward the top. Fine grained facies are sometimes bioturbated (box-work burrows of the ichnogenus *Ophiomorpha* or possibly *Thalassinoides*) and exhibit a macro-fossil association comprising large bivalves, gastropods, rhodoliths, bryozoa, echinoids and rare corals, suggesting a shallow marine to beach environment. The basaltic plateau is composed of thick submarine sheet flows and pillow-lavas (attributed to the lower member of Pico da Antónia Volcanic Complex), whose top forms a regular surface at 175–190 m asl and is covered by later subaerial flows (attributed to the upper member of Pico da Antónia Volcanic Complex). These later flows preserved the flat morphology on the top of the plateau but spilled to lower levels (filling an incipient palaeo-topography imposed on the previous sequence) without reaching the sea (indicating that the relative sea-level was already significantly lower). Above these units it is possible to find phonolitic domes and lavas (still included in the Pico da Antónia volcanic sequence) and younger strombolian cones (attributed to the Assomada/Monte das Vacas Formation) but none show clear submarine morphology. The top of the submarine sheet flows, a very regular feature, can be used as a marker for the relative sea-level at the time of their extrusion.

5.3.4.5 The Assomada/Monte das Vacas Formation(s)

The Assomada/Monte das Vacas Formation seems to have produced exclusively subaerial products, however this unit is only mapped above 100 m asl, with the exception of only two cones whose base is at 40–50 m asl. Thus we infer that the contemporaneous sea-level was generally below 40 m. The age relationship between this unit and the Quaternary sediments is not clear, but both units are probably partially synchronous.

5.3.4.6 Quaternary Sediments and Wave-cut Surfaces

Geomorphological expressions of Quaternary sea-level palaeo-positions are also frequently found in Santiago, on the fringes of the present island edifice. These are mostly represented by marine abrasion surfaces, with occasional boulder accumulations and a few small carbonate terraces. They are not as extensive as in the eastern islands, but are still discernible. Serralheiro [22] describes seven main levels: 2–4, 5–10, 15–25, 30–40, 50–60, 60–80, and 80–100 m asl. For a detailed description of these features we suggest the reading of Serralheiro [22]. We shall mention some of the best places to observe these features.

In the western coast, between Porto Rincão and Ponta da Covinha, the whole sequence was carved during Quaternary times by marine erosion forming a complex geomorphology with evidence for marine abrasion surfaces at 15–25, 30–40, 60–80, and 85–100 m asl [22]. As noted by Serralheiro [22], significant boulder accumulations can be found covering the abrasion surface between 80 and 100 m, and scarce conglomeratic outcrops exist in Porto Rincão between 15 and 25 m.

In the region of Porto Mosquito, a composite marine abrasion surface was cut during different periods of the Quaternary, creating a prominent sea cliff. This surface is presently at 20–40 m asl and probably corresponds to two or three different high stands (it is possible to distinguish two or three different levels in this surface, sometimes marked by boulder accumulations). The backbeach cliff is presently a few hundred meters distant from the sea.

In the eastern coast, sediments can be found at Praia Baixo and Senhora da Luz, or even further south. In Praia Baixo the sediments comprise conglomerates and calcarenites rich in bivalves, gastropods, echinoids and barnicles. They probably correspond to the 15–25 m deposit, however, its base can be found at lower elevations since the deposit can be up to 20 m thick [22]. In Senhora da Luz, several fossiliferous deposits can be found at 15–25 m, 5–10 and at 2–4 m [22].

5.3.5 Fogo

Fogo is the youngest and most prominent island of the archipelago (2,829 m asl), with a record of frequent historical eruptions up to 1995. The island is a polygenetic stratovolcano in a Monte Somma-Vesuvio-like association, probably built entirely during the Quaternary. The young age and the vigorous volcanic activity of the island leads to flow-mantling of most palaeo-markers of relative sea-level change. The only occurrences of submarine volcanism are the Fajãzinha surtseyan cone, built slightly above present sea-level, and a pillow lava sequence exposed in the sea-cliff west of Fajãzinha. The pillows, of probable Pleistocene age, cover beach sand deposits and crop out up to ~ 20 m asl (José Madeira, Personal Communication, 2009).

5.3.6 Brava

Brava is the westernmost island of the southern chain. Its small size contrasts with its high topography, and its morphology is complex, with young volcanic landforms on top of an older edifice.

The volcanostratigraphy of the island was described by Madeira et al. [48] and Mourão et al. [49], and comprises two main units separated by an important unconformity: (1) a lower unit composed of submarine lavas, intruded by an alkaline/carbonatitic complex; and (2) an upper subaerial unit, mainly phonolitic.

According to Madeira et al. [48], the submarine volcanic products of the lower unit comprise pillow-lavas and hyaloclastites that occur, presently, up to ~ 350 m asl. The same authors note that the erosional surface imposed on these lavas is very irregular and is subaerial. The age of these submarine volcanic products is unknown, but their top marks a minimum relative sea-level height for their age.

5.3.7 São Nicolau

São Nicolau is the easternmost island of the northern chain. Its morphology and structure suggest an island formed by composite fissure volcanism along two main rift arms: the most prominent, oriented in an E–W to WNW–ESE direction, and a lesser one, oriented N–S, in the western area of the island. In terms of its volcanostratigraphy, São Nicolau resembles Santiago [22]. São Nicolau volcanism extends from Miocene to Quaternary [23, 25] and the edifice is probably in the early post-erosional stage. The volcanostratigraphy of the island was defined by Macedo et al. [23] and is summarized in Table 5.5, with age and inferred sea-level constraints. The volcanostratigraphy comprises [23] from oldest to youngest: Ancient Eruptive Complex; Late Miocene Marine Sediments; Figueira de Coxe Formation; Conglomerate-Breccia Deposit; Main Eruptive Complex; Preguiça Formation; Monte Gordo Formation; and Quaternary sediments (see Fig. 5.10).

5.3.7.1 The Ancient Eruptive Complex

The Ancient Eruptive Complex of São Nicolau is especially well exposed in the Ribeira Brava valley. This unit is intensely weathered and intruded by an important dyke swarm (up to 90% of the exposed sequence), making it very difficult to observe what was the country rock before the intrusion of the dykes. However, the observable material is intensely altered to a yellow clay of probable palagonitic nature, suggesting it to be of hyaloclastitic nature. The whole unit, including the large majority of the dykes that cut through its sequence, is clearly truncated atop by an important erosive surface preserved by the overlying units. This unconformity is well exposed in the right bank of Ribeira Brava, below the Figueira de Coxe pyroclasts. Thus, the lithologies that compose the Ancient Eruptive Complex are older than any of the dated units by Duprat et al. [25] and remain undated.

A series of eroded phonolitic vents and associated products were also included in the Ancient Eruptive Complex by Macedo et al. [23] and described as subaerial in nature. The presence of possible subaerial products in the sequence of the Ancient Eruptive Complex needs further investigation because of its significance regarding uplift reconstructions.

5.3.7.2 The Monte Focinho Late Miocene Marine Sediments

Shallow marine sediments, attributed to the Late Miocene [50, 51], can be found in Campo da Preguiça separating the Ancient Eruptive Complex from the submarine volcanic products of the Figueira de Coxe Formation. The sediments occur in discontinuous outcrops ranging from 170 to 227 m asl, and overly products of the Ancient Eruptive Complex including a dismantled phonolitic intrusion. Torres and Soares [52], and later Bernoulli et al. [51], proposed that this intrusion corresponds to a crypto-dome who warped the sediment beds when it was emplaced, tilting them to the W. However, we are more inclined towards the interpretations by Serralheiro and Ubaldo [50] and Macedo et al. [23] that consider this intrusion to be older than the sediments and part of the bedrock where the later were deposited.

Table 5.5 Summarized volcanostrigraphy of São Nicolau, with age and sea-level constraints

Units[a]	Main characteristics	Age data for exposed rocks	Evidence for contemporaneous sea-level	Palaeo-depth constraints	Sea-level constraints	Key localities for sea-level markers
Quaternary sediments	Marine terraces (and wave-cut surfaces), alluvium, colluvium, etc	Younger than 1.8 Ma?	Marine terraces and wave-cut surfaces ranging at 5–7, 10, 20–30 and 50–60 m asl	Beach deposits and shallow marine sediments. Inferred palaeo-depth <5 m	Sea-level up to 60 m asl	Carriçal; Coelha
Erosional and conformable contact, depending on localities						
Monte Gordo Fm	Post-erosional subaerial stage with rare submarine morphologies	Younger than 1 Ma (Ar/Ar)[b]	Subaerial flows exist down to 5 m asl	–	Sea-level below 5 m asl	–
Erosional contact						
Preguiça Fm	Post-erosional subaerial stage with extensive coastal lava deltas, especially in the eastern part of the island. Presence of shallow marine sediments at the base of the lava deltas	Between 1.74 ± 0.12 and 1.06 ± 0.07 Ma (Ar/Ar)[b]	Extensive lava delta of Terra Chã with the passage zone at 40 m asl. Other lava deltas with passage zones at lower elevations	Passage zone clearly visible at 40 m asl	Sea-level up to 40 m asl	Terra Chã; Carriçal; Preguiça
Erosional contact						
Main Eruptive Complex	Main shield-building stage, mostly comprising effusive products with a fringe of submarine morphologies on the periphery of the volcanic edifice. Marine sediments can be found within the submarine lavas	Between 4.73 ± 0.28 and 2.68 ± 0.02 Ma (Ar/Ar)[b]	Passage zone is clearly visible in the eastern part of the island very regularly at 100 m asl. Locally lower in other parts of the island	Passage zone clearly visible at 100 m asl	Sea-level up to 100 m asl	Eastern part of the island; Turil

(continued)

Table 5.5 (continued)

Units[a]	Main characteristics	Age data for exposed rocks	Evidence for contemporaneous sea-level	Palaeo-depth constraints	Sea-level constraints	Key localities for sea-level markers
Erosional and conformable contact, depending on localities						
Conglomerate-breccia deposit	Conglomerate deposit: quiescence and erosion[a]	Unknown	Unknown	Unknown	Unknown	–
Erosional contact						
Figueira de Coxe Fm	Exclusively submarine lavas.[a] Unit deeply eroded. Possibly early submarine shield-building stage	Between 6.18 ± 0.89 and 5.73 ± 0.12 Ma (Ar/Ar)[b]	Entire unit is submarine, cropping out up to 270 m asl	Top is eroded. Elevation of highest point assumed as minimum sea-level height	Sea-level in excess of 270 m asl	Figueira de Coxe, Campo da Preguia; Queimada
Conformable contact with Figueira de Coxe Fm						
Late Miocene marine sediments	Shallow marine sediments[a,c,d]	Between 11.8 and 5.8 Ma or possibly between 6.2 and 5.8 Ma (Biostratigraphy)[d]	Sediments of marine origin, cropping out up to 227 m asl	Between 60 and 20 m, according with the morphology and fossil content of the deposit	Sea-level in excess of 250 m asl	Campo da Preguia
Erosional contact						
Ancient Eruptive Complex	Severely altered volcanic products of possible submarine origin, intensely intruded by a dyke swarm.[a] Unit deeply eroded. End of seamount stage?	Pre-Late Miocene?	Entire unit is possibly submarine	Unknown	Unknown	–

[a] Macedo et al. [23]
[b] Duprat et al. [25]
[c] Serralheiro and Ubaldo [50]
[d] Bernoulli et al. [51]

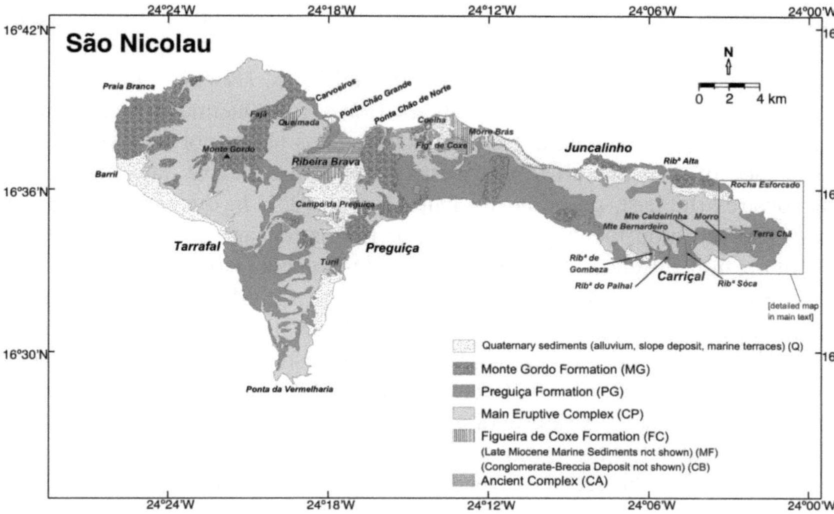

Fig. 5.10 Geological map of São Nicolau, modified from Macedo et al. [23]

Our opinion is based in the fact that no baking or peperitic structures can be seen in the sediments near the contact, as one would expect if the contact was intrusive in nature.

The marine sediments seems to correspond to an age comprehended between 11.8 and 5.8 Ma or even between 6.2 and 5.8 [51], and vary from friable calcarenites in the topographically higher outcrops (at Monte Focinho) to more compact bioclastic calcarenites in the lower outcrops. The friable calcarenites are very rich in bivalves, gastropods, and echinoid spicules, and probably represent beach deposits. Deformation marks are visible at the top due to the emplacement of the Figueira de Coxe pillow-lavas on top of the soft sediments [23, 50]. This interface is presently at a maximum elevation of ~ 227 m asl, and in the vicinity these lavas reach ~ 250 m asl. The lower outcrops exhibit more compact, fossil-poor bioclastic facies with an association mostly comprising clasts of rhodoliths and larger bivalves, suggesting slightly deeper water facies. We interpret this lateral variation as a normal distribution of facies in depth over a pre-existing irregular topography. We therefore infer a minimum contemporaneous relative sea-level height between 250 and 270 m asl.

5.3.7.3 The Figueira de Coxe Formation

The Figueira de Coxe Formation is exclusively composed of submarine lavas resting on top of the Ancient Eruptive Complex and the marine sediments [23]. This unit was originally described in the Figueira de Coxe region (from where the formation takes its name) where they are best preserved and reach higher elevations, cropping out from the present sea-level up to elevations close to 260 m asl. The unit is eroded and loses elevation towards Juncalinho in the east, being covered by the Main Eruptive Complex subaerial lavas. However, this formation also occurs in Campo de Preguiça, Ribeira Brava and Queimada regions where they reach similar elevations (above 200 m asl). Thus we do not see evidences for the existence of a N–S fault to the west of Figueira de Coxe, and the possible tilting towards the east, proposed by Duprat et al. [25]. In our view, the irregularities observed atop the Figueira de Coxe Formation are due to a subsequent period of erosion correlated to the Conglomerate-Breccia Deposit of Macedo et al. [23] and the marine sediments found at the base of the Main Eruptive Complex (well exposed in the eastern part of the island). Hence, the position of the Figueira de Coxe Formation outcrops reinforces the suggestion of a contemporaneous minimum relative sea-level height between 250 and 270 m above the present sea-level. The Figueira de Coxe Formation corresponds to Stage 1 defined by Duprat et al. [25], attending to the locations and stratigraphical position of the samples collected by these authors, as described in their paper. Consequently, the proposed age interval for the extrusion of this formation is between 6.2 and 5.7 Ma [25].

The products of this unit mostly comprise pillow-lavas and pillow and hyaloclastite breccias. However, in Ribeira Brava is composed of submarine pyroclasts [23] suggesting the presence of an important eruptive centre in this place.

5.3.7.4 The Main Eruptive Complex

Unconformably above the previous units lies the Main Eruptive Complex. This unit corresponds to the main shield-building stage of São Nicolau, and constitutes the bulk of the exposed edifice. This unit seems to correspond to Stage 2 defined by Duprat et al. [25], attending to the locations and stratigraphical position of the samples collected by these authors, as described in their paper. Thus, it was probably extruded between 4.7 and 2.6 Ma [25].

The unit mainly comprises effusive products that reached the sea, thus creating a fringe of submarine morphology on the periphery of the volcanic edifice. The passage zone between the subaerial and submarine morphologies can be best observed in the eastern part of the island, where it can be found very regularly at 100 m asl (see Fig. 5.11). We thus infer the contemporaneous sea-level at 100 m asl. In the central and western part of the island, the presence of young flows mask the boundary between subaerial and submarine morphologies, making it difficult to observe, or in some cases the contact is erosive in nature instead of continuous. Notwithstanding, submarine flows can be found up to 90 m asl. The presence of marine sediments is frequent at the base of the sequence or intercalated in the submarine lavas.

The easternmost part of the island constitutes an important case-study of relative sea-level change (see Figs.5.11, 5.12, 5.13). Here a headland of the Main Eruptive Complex was "fossilized" by the later flows that formed the extensive lava delta of Terra Chã. The shape of this palaeo-headland can be deduced by the

Fig. 5.11 Photographs of examples of palaeo-markers of sea-level in the Cape Verdes. **a** Main Eruptive Complex sequence in eastern São Nicolau (Boca da Ribeira-Ponta de Tope de Capa and Castilhiano region, seen from the south), where the passage zone between the subaerial lavas (CP) and submarine lavas (CPsub) is clearly visible at ~ 100 m asl. **b** Passage zone between Main Eruptive Complex subaerial lavas and Main Eruptive Complex pillow-lavas and hyaloclastites, at ~ 100 m, in Ribeira António Barbacante, eastern São Nicolau. **c** Lava delta of Terra Chã, in eastern São Nicolau, probably belonging to the Preguiça Fm and overlying a palaeo-coast carved in the Main Eruptive Complex (marked by a *dashed line*). **d** Detail of the contact between the sequence of the Main Eruptive Complex and the overlying sequence of the Terra Chã lava delta (in Talhada, eastern São Nicolau). The Main Eruptive Complex sequence comprise pillow-lavas and hyaloclastites (CPsub) at the base and subaerial lavas (CP) on top, with a passage zone at ~ 100 m asl. The Terra Chã volcanic sequence overly the previous sequence, with calcarenites (calc) along the contact, and exhibit a passage zone between the subaerial (PG) and submarine (PGsub) lavas at ~ 40 m asl

~ 60 m geomorphological step that runs from Talhada to Ribeira de Castilhiano, separating the gently dipping higher elevations west of this line from the flat surface of the Terra Chã plateau to the east. The young lava flows covered the Main Complex sequence, in-filled the existing east-draining valleys, and spilled out to form the Terra Chã volcanic delta, extending the island to the east. The unconformity thus created, and the sequence of both the underlying and overlying units, can be best observed in the valleys of Ribeira de Castilhiano, Ribeira de António Barbacante, Covoada de Bodela, Ribeira de Falcão (an extension of the previous two confluent valleys), and in the southeast coast between Ponta Barroso and Ponta Talhada. These flows probably originated in a WNW-ESE lineament of vents that include Monte Caldeirinha and Morro cones.

The headland sequence comprises, from the base to the top: Ancient Eruptive Complex (?); Marine carbonates; Main Eruptive Complex submarine lower

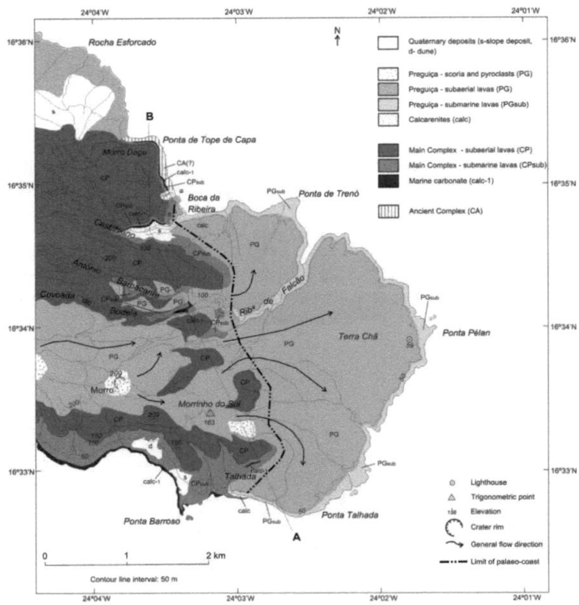

Fig. 5.12 Geological map of the easternmost region of São Nicolau (see Fig. 5.11 for photographs)

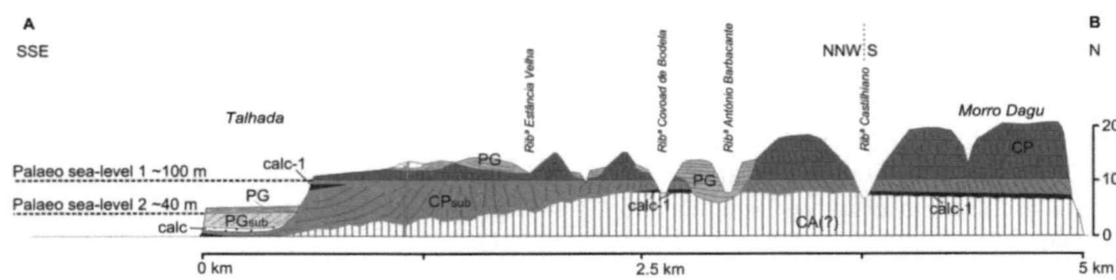

Fig. 5.13 Schematic cross section of the easternmost region of São Nicolau. Inferred local palaeo sea-levels are shown. Key as map (see Fig. 5.11 for photographs)

member; and Main Eruptive Complex subaerial upper member. In this area the passage zone between the submarine and subaerial members of the Main Eruptive Complex can be found regularly at ~100 m asl (in Ribeira de Castilhiano, Ribeira de António Barbacante, Ribeira de Covoada de Bodela and Talhada).

The Castilhiano valley provides a good cross-section of the headland sequence, despite its lower slopes being covered by scree. The bottom of the valley is carved in the Ancient Eruptive Complex (?) and comprises intensively weathered tuffs (submarine?) intruded by steep dipping dykes. It crops out up to ~80 m asl. The marine carbonates reach a thickness up to 6 m and lie unconformably over the previous unit. The carbonates are characterised by a framework of rhodoliths, enveloped in a fine bioclastic grainstone matrix. Fossils of echinoids and bivalves are also present. By comparison with other present and past environments [53, 54], we suggest these sediments correspond to a medium energy shelf environment.

The assemblage of echinoid/rhodoliths, together with the indications of a medium energy environment, suggests a palaeo-depth ranging 10–50 m. Above the sediments, up to 102 m asl, the Main Complex submarine sequence are composed of a 3 m thick submarine sheet flow immediately overlying the carbonates, and hyaloclastite and pillow breccia above it. Finally, atop the pile, the subhorizontal Main Complex subaerial flows crop out. The whole sequence is also superbly exposed on the coast running from Boca da Ribeira to Ponta de Tope de Capa. Very similar lateral equivalents of the previously described carbonates can also be found in the adjacent valley of Covoada de Bodela and on the coast in Talhada-Ponta Barroso.

In the Talhada region the observable sequence comprises: basal outcrops of Ancient Eruptive Complex (?) tuffs and lavas (up to 20 m asl); compact marine carbonates lying unconformably over the previous unit and up to 5 m thick, with occasional individual pillows included in the sediments as clasts; Main Complex submarine flows from 5–20 m asl up to $\sim$100 m asl; and Main Eruptive Complex subaerial flows. Close to the top of the submarine sequence, a 1 m thick well-cemented calcarenite bed, rich in bivalves, corals, rhodoliths and echinoids, can be found within the pillow-lavas.

Further west in Carriçal area, the Main Eruptive Complex sequence is composed of a submarine lower member of seaward dipping foresets of hyaloclastite breccia and pillows and an upper member of subaerial subhorizontal lavas. The passage zone is best observed in the Ribeira do Palhal valley, 800 m upstream from the village of Carriçal. In this location it is found at an elevation of $\sim$70 m asl, suggesting a variation in sea-level height during the formation of the Main Complex.

In the central and western part of the island the Main Complex submarine lower member is usually covered by young flows and the passage zone is difficult to observe. We studied the sequence in Traje Ribeirão de Cruz, Rib. da Ponta da Pataca and Turil area (to the south of Preguiça) where the top of the submarine flows reaches $\sim$90 m asl. These outcrops reveal a sequence composed of submarine lavas (pillows and sheet flows) with occasional lenses of marine carbonate beds. The carbonates are mostly unbedded calcarenites, conglomeratic at the base, and rich in bivalves, gastropods, bryozoan, and sporadic corals. The maximum thickness of these lenses is between 12 and 15 m [23]. The sediments are exhibit a facies compatible with transition zone/offshore deposits. This constitutes further evidence for a relative sea-level height close to 100 m during the extrusion of the Main Eruptive Complex.

5.3.7.5 The Preguiça and Monte Gordo Formations

The Preguiça and Monte Gordo Formations seem to constitute two post-shield volcanic units that in-filled palaeo-topographies including marine abrasion surfaces. These two formations probably correspond respectively to Stage 3 and 4 of Duprat et al. [25].

The Preguiça volcanic activity probably took place between 1.7 and 1 Ma [25] and created prominent coastal lava deltas like the Terra Chã and Carriçal structures, the Ponta Chão Grande to Carvoeiros platform—all features previously attributed to the Main Complex by Macedo et al. [23]—and possibly the delta to the southeast of Tarrafal (see electronic supplement for details).

The Terra Chã effusive delta is the most important sea-level marker of the Preguiça Formation (see Fig. 5.11). It comprises a lower submarine member mainly composed of steeply eastward-dipping breccia and pillow foresets covered by flat lying subaerial flows. A discontinuous calcarenite bed of a few meters thick is occasionally present underneath the sequence, lying unconformably over the older formations. The passage zone in the delta sequence is found very regularly at $\sim$40 m asl. In Boca da Ribeira (Castilhiano), the submarine flows overlie a 1–3 m thick conglomerate (with calcarenitic matrix), very rich in large oysters. Boulders of pillow-lavas are present in the deposit. This deposit, in turn, unconformably overlies intensively weathered tuffs possibly belonging to the Ancient Eruptive Complex (?). The young lavas were considered to be part of the Main Eruptive Complex by Macedo et al. [23], but they must be considered as part of the younger post-erosional Preguiça Formation, due to their field relations and geochronological data. Radiometric ages determined by Duprat et al. [25] in the young flows at Ribeira Falcão yielded ages between 1.7 and 1.15 Ma.

Further west in Carriçal area it is possible to observe a marine abrasion surface carved in the Main Complex, covered by young subaerial lava flows that we group with the Preguiça Formation. The abrasion surface, extending 600 m inland, was cut in the submarine lavas and it is currently at 15–35 m asl. It is covered by a thin sequence of subaerial lava flows that spilled from the proto-valleys of Ribeira do Palhal and Ribeira de Gombeza. The vents that originated these flows were not identified. In places it is possible to find a thin conglomeratic deposit between the two sequences. Lava flows from the Monte Bernardeiro and Monte Caldeirinha cones were also erupted around the same time, creating the recognisable morphology of Carriçal effusive fan probably around 1 Ma, the age determined by Duprat et al. [25] in Ribeira de Sóca (see Fig. 4.6). The flows from Ribeira do Palhal reached the contemporaneous shore and covered the fossiliferous estuarine deposits of the same creek. This suggests that the coastline, and that the sea-level height were similar to the present day's. The sequence, including the estuarine sediments, can be observed in Ribeira do Palhal river mouth, near Carriçal. These sediments comprise an 8 m thick series of imbricated conglomerates, intercalated with calcarenites/sandstones with cross-lamination, rich in bivalves and gastropods. The young subaerial lavas were in turn submerged and partially eroded to create another marine abrasion surface, presently at 20–30 m asl and marked by the presence of scattered outcrops of shallow marine carbonates. These are conglomeratic and rich in fossils of corals, bivalves, gastropods and echinoids, an association typical of a high energy environment. A small cliff, etched in the young flows during this period, can be seen 300 m to the east of Carriçal village (see Fig. 4.6).

The lava delta located to the southeast of Tarrafal, possibly belonging to the Preguiça Formation, exhibits a submarine morphology whose passage zone can be presently seen no higher than 5 m asl.

The Monte Gordo Formation corresponds to the majority of the well preserved cones in São Nicolau, and associated products (like, for example, the Fajã flows, wrongly attributed to the Preguiça Formation by Macedo et al. [23]. Because many of the products and morphologies of the Preguiça Formation are still well preserved, it is very difficult to distinguish them from the younger Monte Gordo Formation. Thus, with a few exceptions, the separation between these units based only in a geomorphological character is somewhat artificial. Notwithstanding, none of the structures attributed to the Monte Gordo Formation seems to exhibit clear submarine morphologies. The Monte Gordo Formation was extruded <1 Ma [25].

5.3.7.6 Quaternary Sediments and Wave-cut Surfaces

A close look at the coastal morphology of São Nicolau reveals the presence of large exposed recent marine abrasion surface(s) up to 50–60 m asl, partially covered by later volcanism and alluvial fan deposits. We shall describe a few areas where these features are well observed.

In the stretch of coast running between Praia Branca and Tarrafal, a flat morphology can be found from the present coastline up to 1.5 km inland. This platform is comprehended between the present sea-level and 50–60 m asl, running against the old backbeach cliff. Between Praia Branca and Barril, the platform is covered by two recent strombolian cones and associated products, probably belonging to the Monte Gordo Formation. Between Barril and Tarrafal, the platform is covered in extensive alluvial fan deposits.

In Ponta da Vermelharia, the platform occurs very regularly up to 200 m inland and up to 20 m asl. Further north and towards Turil, the platform widens up to 1 km inland and is extensively covered in alluvial fan deposit. However, scattered outcrops of fossiliferous calcarenites can be found up to 30 m asl.

Along the coast between Morre Brás and Ponta Chã de Norte, the platform is also very wide and several outcrops of fossiliferous calcarenites can be found at 20–30, 10 and 5–7 m asl. The deposits in Coelha region (20–30 m asl) deserve a special mention due to quality of their fossil preservation and because they are covered by pyroclasts of the Monte Gordo Formation [23]. These sediments comprise conglomerates with calcarenite matrix, not exceeding a total thickness of 2 m, and their fossil association includes large oysters, other bivalves, gastropods, corals, bryozoa and very well preserved worm-burrows.

The platform is also evident along the coast stretching from Juncalinho and Rocha Esforcado, where a palaeo coast line, defined at he base of the backbeach cliff, is very evident at 30 m asl.

5.3.8 Santa Luzia and Islets

The geology of Santa Luzia, Ilhéu Branco and Ilhéu Raso is almost unknown. Regarding palaeo-markers of sea-level, brief mentions by Bebiano [37] and by

Torres and Soares [52] include raised Quaternary beach deposits, but localities and precise elevations are very vague. Recent workers report a (Quaternary?) lava delta whose passage zone is between 1.5 and 3 m asl, and Quaternary terraces (consolidated beach deposits) at 0, 3–5, 12–15 and 21–26 m (Brum da Silveira, Personal Communication, 2009).

5.3.9 São Vicente

São Vicente seems to be a deeply eroded composite volcano in a rejuvenated stage. The morphology of the edifice is poorly preserved but it is still possible to deduce some features, like the sea-invaded partially-destroyed caldera that constitutes the Mindelo bay [55], [22]. The volcanostratigraphy of the island was defined by Serralheiro [22] and is summarised in Table 5.6. Later, radiometric studies at extrusive rocks, by Jørgensen and Holm [56] and Holm et al. [21] reported ages spanning between 6.6 and 0.3 Ma, divided into three main periods of activity (see Table 5.6, Fig. 5.14).

In striking contrast with the majority of the other islands in the archipelago, the exposed sequence of São Vicente seems to be almost devoid of submarine lithologies. The only scarce evidence for relative sea-levels higher than the present level are a few Quaternary terraces found scattered around the island, and some possible marine abrasion surfaces.

The main Quaternary marine deposits, some of which were already identified by Bebiano [37], are located in São Pedro (1 km to NE of the village), Salamansa-Baia das Gatas, and Curral de João Paula areas.

The marine sediments near São Pedro crop out 1 km to NE of the village, at $\sim$18 m asl. These are conglomerates to micro-conglomerate beds, comprising terrigenous pebbles in a calcarenitic matrix, up to 1 m of total thickness (normally less). Abundant bivalves and gastropods can be found. The deposit indicates a shallow marine/beach depositional environment that marks a palaeo-shore.

Around the coast of Salamansa/Baia das Gatas, at 2–4 m asl, there is a raised beach deposit atop the young lavas that constitute this peninsula. They form a discontinuous belt parallel to the coastline and comprise very fossiliferous bivalve- and gastropod-rich calcarenites, probably formed during the last highstand.

Finally, in Ribeira de João Paula, on top of the young flows originated in the Vulcão Viana and the Curral de João Paula cones, a calcarenite to micro-conglomerate deposit crops out in the riverbed 12 m asl and is up to 70 cm thick. The base of the deposit is micro-conglomeratic, rich in volcaniclastic material and shell debris, and at the top grades to a calcarenite with frequent oysters and other bivalves, gastropods, and corals. The deposit is extensively covered by an alluvial fan deposit and is only visible in the riverbed where washouts clear the recent cover. The macro-fauna indicates deposition in a shallow marine environment, suggesting a contemporaneous sea-level height of $\sim$15 m asl. A minimum age of

Table 5.6 Summarized volcanostrigraphy of São Vicente, with age and sea-level constraints

Units[a]	Main characteristics	Age data for exposed rocks	Evidence for contemporaneous sea-level	Palaeo-depth constraints	Sea-level constraints	Key localities for sea-level markers
Quaternary sediments	Marine terraces, alluvium, colluvium, etc	Younger than 0.33 Ma?	Marine terraces and wave-cut surfaces ranging at 2–4, 12 and 18 m asl	Beach deposits and shallow marine sediments. Inferred palaeo-depth <5 m	Sea-level up to 23 m asl	São Pedro; Salamansa-Baía das Gatas; Curral de João Paula
Erosional and conformable contact, depending on localities						
Recent volcanic products	Post-erosional stage: strombolian-hawaiian and tuff (phreato-magmatic) cones and associated lava flows. Equivalent to recent volcanics[b]	0.33 ± 0.05 Ma (Ar/Ar)[b]	Unit possibly entirely subaerial, cropping down to present sea-level	Phreato-magmatic cone may indicate proximity to sea-level	Sea-level at or slightly below present level	Curral de João Paula
Erosional contact						
Monte Verde Eruptive Complex	Effusive and explosive subaerial basalts and phonolites	Between 4.73 ± 0.08 and 4.43 ± 0.04 Ma (Ar/Ar)[b]	Subaerial flows down to present sea-level	–	Sea-level at or below present level	–
Erosional and conformable contact, depending on localities						
Conglomerate-breccia deposit	Conglomerate deposit: quiescence and erosion[a,d]	Unknown	Unknown	Unknown	Unknown	–

(continued)

Table 5.6 (continued)

Units[a]	Main characteristics	Age data for exposed rocks	Evidence for contemporaneous sea-level	Palaeo-depth constraints	Sea-level constraints	Key localities for sea-level markers
Erosional contact						
Ancient Eruptive Complex	Extrusive basaltic, phonolitic and carbonatitic rocks, intensely intruded by large plutons of syenites and gabbros.[a] Probably equivalent to the Old Eruptive Complex[b]	Between 6.6 ± 0.3 and 5.68 ± 0.22 Ma (Ar/Ar)[b,c]	Unknown	–	Unknown	–

[a] Serralheiro [22]
[b] Jørgensen and Holm [56]
[c] Holm et al. [21]
[d] Serralheiro [55]

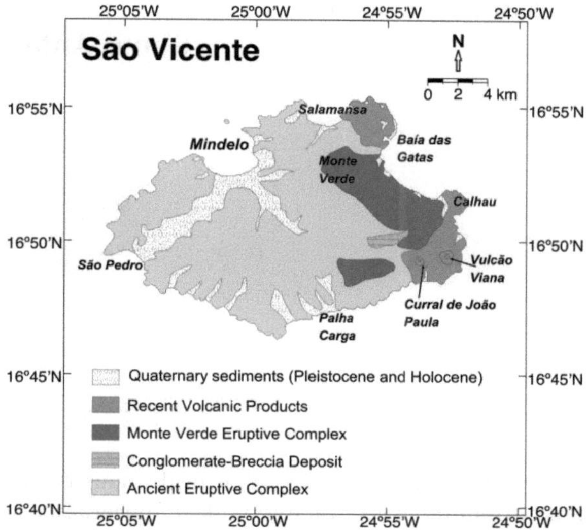

Fig. 5.14 Geological map of São Vicente, based on the volcanostratigraphy defined by Serralheiro et al. [40] and Jørgensen and Holm [56]

~0.3 Ma is inferred, because this age is attributed to the lavas that underlie the deposit. Similar deposits are reported by Bebiano [37] in Palha Carga valley.

5.3.10 Santo Antão

Santo Antão is one of the largest and most prominent islands of the Cape Verde Archipelago. This island is an elongated shield volcano possibly in an early post-erosional stage, created by fissural eruptions along a main NE–SW rift zone. The edifice is cut by deep valleys and canyons, but the shape of the shield volcano that constitutes the bulk of the sequence is still perceptible. Later post-erosional eruptions in-filled the valleys and scattered the island with cones, one of which forms the Tope de Coroa strato-volcano, the highest peak in the island (1979 m asl) (Fig. 5.15).

Despite the work of Plesner et al. [26], the detailed stratigraphy of Santo Antão is still not well known. These authors, though, carried out a geochronological study that revealed volcanic activity from 7.6 Ma until 0.09 Ma BP.

Santo Antão is also almost devoid of submarine lithologies. The only submarine lavas reported to date on the island are located in Ribeira Grande and in Ponta do Atum (Tarrafal) areas. Marine carbonates are also very rare and seem to be restricted to a few Quaternary marine terraces of small extent in Ponta do Atum and Ponta do Sol, as mentioned originally by Bebiano [37]. The Ribeira Grande pillow-lavas can be found up to 20 m asl but are probably fault-bounded [21] and thus regarded as a poor tracer of sea-level.

The Ponta do Atum sequence constitutes a solid tracer of sea-level, but its young age only pinpoints a few Quaternary sea-level positions (see Fig. 5.16). The

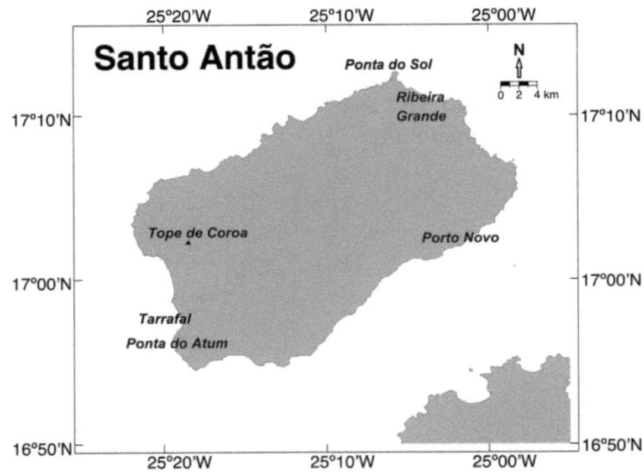

Fig. 5.15 Location map of Santo Antão

sequence is exposed along >1 m in the southern part of Tarrafal bay, and comprises from the base to the top: older subaerial lavas; coastal sediments; younger subaerial lavas that grade laterally to submarine (at lower elevations); Pleistocene beach deposits; Holocene beach deposits. The older subaerial lavas are of unknown age and form a sequence of southwest dipping flows in which an abrasion surface and cliff were carved. This surface subsequently filled by >40 m of coastal sediments comprising terrigenous conglomerates and sands, with fossiliferous calcarenites at the base. The conglomerates exhibit poorly sorted rounded clasts, in a sandy matrix. This sedimentary sequence seems to correspond to a coastal fan-delta deposit, as suggested by the transition from a shallow marine/beach-type facies to a more debris-flow facies. The sediments were then covered by lavas that flowed from the NE, probably from one of the many vents located in the plateau above. The lavas flowed over the sediments, baking the topmost layer, and passed laterally into pillow-lavas and submarine sheet flows when they entered the sea. The passage zone is irregular and difficult to detect but seems to be presently at 12–15 m, or possibly a few meters higher, suggesting a contemporaneous relative sea-level at this elevation. The young lavas probably correspond to the Young Tarrafal volcanic unit identified by Holm et al. [57], which erupted between 0.4 and 0.2 Ma. The whole sequence was again eroded more recently, leading to the creation of a marine abrasion surface presently between 6 and 12 m. This surface exhibits scattered coarse calcarenites with well preserved corals, gastropods, oysters and other bivalves (in the palaeo shoreface), and the contemporaneous foreshore is marked by the presence of well preserved echinoderm holes dug in the volcanic bedrock. A linear accumulation of beach boulders, parallel to the coastline and interpreted as Holocene storm deposits, marks the top of the present day foreshore.

The Ponta do Sol carbonate sediments are thin (20–30 cm) calcarenites, with abundant bivalves and gastropods, revealing a shallow marine origin, deposited over conglomerates and covered by debris flow deposits. Presently this terrace is reduced to a few small outcrops at 15 m asl.

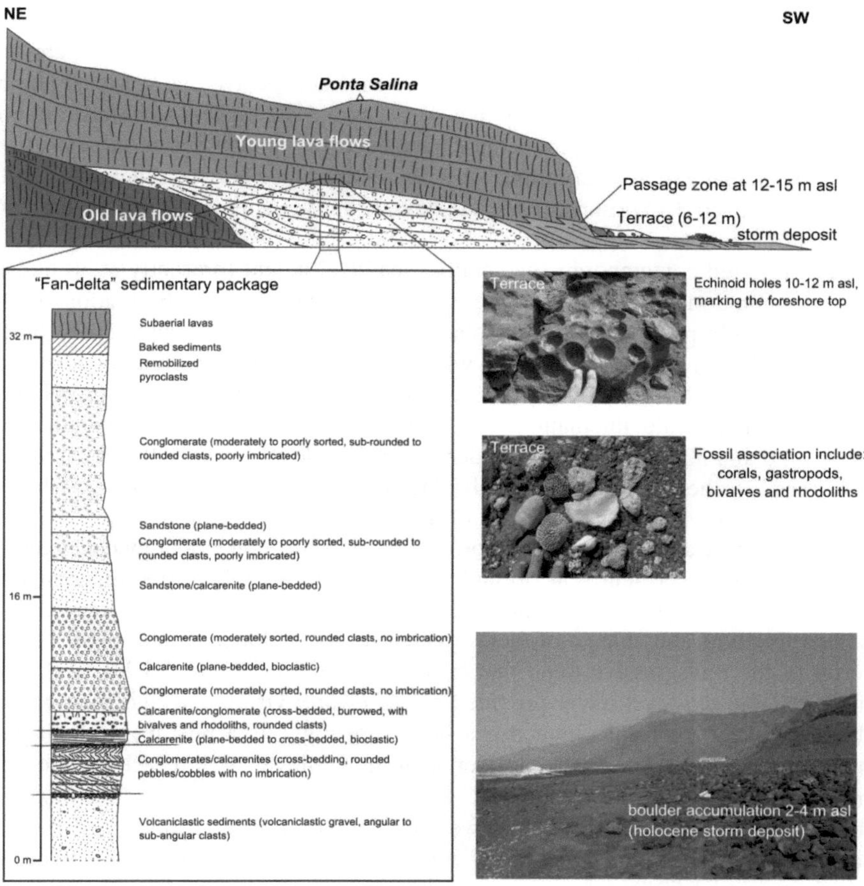

Fig. 5.16 Schematic cross section of Ponta do Atum (Santo Antão), and details

5.4 Inferences of Uplift and Subsidence

5.4.1 Sal

The nature of the Ancient Eruptive Complex, almost exclusively submarine, seems to support the hypothesis that this unit corresponds to the final stages of the seamount phase. This hypothesis is also supported by the presence of micritic carbonates within the submarine sequence. Due to erosion it is impossible to quantify what the original height of the submarine volcanic pile was, and thus to quantify the uplift. The poorly constrained age reported for the volcanic pile is also a barrier to solid event reconstructions. Despite this problem, we can use the intercalated carbonates and its inferred age and depth to estimate the vertical displacement. The foraminiferal content present in these sediments suggest a depositional depth below 300 m and an Early to Mid Miocene age. We therefore infer a vertical displacement greater than 350 m, since the average sea-level height between 21.5 and 14.5 Ma corresponds to -2 m and the highest seastands did not exceed 17 m asl [27]. This uplift must have occurred mostly before the extrusion of the Principal Eruptive Eruption, since the discordance between these two units suggests subaerial exposure of the Ancient Eruptive Complex prior to the formation of the latter.

The inferred contemporaneous relative sea-level height (presently at 50 m asl) for the submarine flows at the base of the Principal Eruptive Formation might suggest a slight positive vertical displacement. The mean sea-level around 16 Ma was very close the present sea-level, with the closest highest seastand at 17 m asl, at 15.6 Ma (see Miller et al. [27]). The Ponta do Altar-Baleia Complex, probably suggests small scale fluctuations in sea-level height, whose markers are now between 0 and 32 m asl. If we consider a mean sea-level between 9 and 11 Ma of 6 m asl [27], the resulting vertical displacement is negligible, especially when a highstand up to 40 m asl is inferred to have occurred at 9.02 Ma [27].

The Fragata and Serra Negra Formations indicate contemporaneous relative sea-level heights presently at 50 and 90 m asl, respectively. This apparent difference in the relative sea-level height between these two consecutive sequences is compatible with the change of facies evident in the sediments, and might be the result of the transgressive trend that occurred between 5.7 and 5.3 Ma [27]. The more reliable sea-level tracer provided by the clearly visible passage zone within the Serra Negra volcanic sequence at 90 m asl suggests a positive vertical displacement of ~ 80 m, if we consider a mean sea-level of 8 m asl for the age interval provided by the radiometric age and its uncertainty of 5.43 ± 0.05 Ma determined by Holm et al. [21]. If we consider the maximum highstand for the same interval, 38 m at 5.475 Ma [27], the vertical displacement for this tracer is still in excess of 50 m.

The series of Quaternary marine terraces and marine abrasion surfaces that cover a large portion of Sal's surface is evidence that the island was partially submerged several times during the Quaternary, creating an extensive shallow water wave-cut platform where sedimentation occurred. During these periods, the island could probably be described as a partially sunken island or incipient guyot, with a few remaining residual volcanic prominences rimmed by shallow marine platforms.

The terraces can be found up to 100 m (Serra Negra), and despite the lack of solid numerical ages for the higher deposits, and using the estimated elevation for the highest seastands in the Quaternary [27], we agree with the hypothesis formulated by Zazo et al. [36] for an uplift trend during the Early Quaternary. This uplift accounts, at least partially, for the vertical displacements inferred for the Ancient Complex, Fragata and Serra Negra Formations but implies that prior subsidence is necessary to explain the position of Ponta do Altar-Baleia and the Main Eruptive Formation. Thus, to reconcile inferred vertical displacements a complex history of uplift and subsidence episodes is required.

5.4.2 Boa Vista

The complete lack of age constraints for the rocks of Boa Vista makes it difficult to constrain the vertical movements that might have affected this island. Despite this, the presence of widespread well-developed Quaternary terraces suggests the island was an incipient guyot several times during the Quaternary. The position of these terraces, up to 105 or possibly up to 130 m asl, suggest some recent uplift since the maximum highstand for the Quaternary did not pass ~24 m asl (see Miller et al. [27]). This suggests a minimum positive vertical displacement of 75 m.

The highest elevation (30–35 m asl) found for any of the submarine facies of the pre-Quaternary formations of the island seems to be within the range of the eustatic variations that dominated the Mio-Pliocene. However, since we infer a Quaternary uplift of at least 75 m, prior subsidence is necessary to explain the position of the older palaeo-markers within the upper limit of the Mio-Pliocene eustatic range.

5.4.3 Maio

The existence of extensive Quaternary marine terraces and abrasion surfaces affecting a large portion of the island's shoreline suggests the island was also an incipient guyot during the Quaternary. The position of these deposits and morphologies up to 100 m asl suggests recent uplift of the same order as that suggested for Boa Vista and Sal.

Since the majority of the volcanic units (Miocene) exhibit exclusively subaerial morphologies down to the present sea-level it is possible to infer that the island did not suffer long-term vertical movements resulting in considerable uplift (after the emplacement of the Mesozoic Basement Complex). In fact, if we take the recent uplift into account, prior subsidence is necessary to place the older palaeo-markers close to or below the present sea-level since their resulting vertical displacement is somewhat below than the minimum Quaternary displacement of 75 m. For example, the Casas Velhas Formation has a passage zone between subaerial/submarine facies at 35–40 m asl. The age of this volcanism, 9.8 ± 0.8 Ma [20], sets sea-level at −4 m asl [27], suggesting a resulting vertical displacement of only ~45 m and not the 75 m expected. Even greater subsidence is necessary if we consider that this passage zone might have been formed during the 9.02 Ma highstand at 40 m [27].

The Mesozoic Basement Complex in Maio is one of the few documented sea-floor uplifts in the archipelago. Regardless of the mechanism responsible for this upheaval, it must encompass a ~4000 m displacement from the original position, or ~6000 m if we consider the swell development to be contemporaneous.

5.4.4 Santiago

The Ancient Complex in Santiago comprises several lithologies of submarine and subaerial nature, of poorly known or unknown age. The relationship between these rocks is not firmly established, nor is the mechanism that led to the existence of portions of uplifted sea-floor in the Ancient Complex sequence. Hence, it is not yet possible to make firm conclusions about the unit's early vertical movements.

The presence of the Flamengos submarine lavas up to ~450 m asl suggests a considerable uplift, as previous authors [21] concluded. We calculate a vertical displacement in excess of 420–450 m, since the eustatic curve (between 4.0 and 5.0 Ma) exhibits a mean sea-level height close to the present level, and a maximum of just 21 m (see Miller et al. [27]). This displacement is probably underestimated because the top of the Flamengos Formation is eroded. Nevertheless it constitutes evidence for one of the highest vertical displacements that can be inferred from the surface geology of the archipelago (excepting Maio's uplifted sea-floor).

If we consider that: (1) the Pico da Antónia Volcanic Complex exhibits submarine morphologies at elevations ranging from a few meters above the present sea-level up to ~270 m asl; (2) the eustatic curve range for the Pico da Antónia age interval (between 2.2 and 3.3 Ma), is ~82 m and its maximum only reaches 25 m asl (see Miller et al. [27]); it is possible to infer that an uplift in excess of 200 m is necessary to explain these disparities. This uplift must have occurred during volcanism. Our estimate of >200 m is considerably higher than previous estimations of >100 m by Holm et al. [21]. However, since this unit is probably constituted by two distinct members and the highest palaeo-markers mostly correspond to lavas of the lower member, a more detailed geochronological study would be necessary to better constrain the timing of uplift since the majority of this uplift might have occurred before the extrusion of the upper member.

The existence of Quaternary abrasion surfaces and marine terraces up to 100 m asl in Santiago is also an indication of probable recent uplift if we consider the maximum height of sea-level in the Quaternary.

5.4.5 Fogo

The only palaeo-markers of sea-level found on Fogo are within the eustatic range for the Quaternary [27]. Hence it seems the island has not experience any significant uplift.

5.4.6 Brava

Despite the lack of ages for the (submarine) lower unit of Brava, its elevation up to 350 m asl is a solid relative sea-level marker. If we take into account that, according to the eustatic curve of Miller et al. [27], there were no highstands above ~50 m asl in the last 26 Ma (the supposed age of the archipelago [19]), it is possible to denote a final positive vertical displacement for this unit of at least 300 m. It is still unknown whether this uplift was related to the intrusion of the alkaline/carbonatite pluton, or if it is related to another process.

5.4.7 São Nicolau

The existence of marine sediments deposited between 11.8 and 5.8 Ma, or even between 6.2 and 5.8 [51], and the existence of the entirely submarine volcanic unit of Figueira de Coxe (6.2–5.7 Ma [25]) up to elevations exceeding 250 m, is a limiting constraint on the relative sea-level height during the period. Miller et al. [27]'s eustatic curve indicates a mean sea-level during the period 6.2 and 5.7 Ma ~10 m below the present level, and a maximum highstand of ~26 m asl at 6.06 Ma. Thus, we can infer a positive vertical displacement ranging from 230 to 280 m for these rocks.

The existence of a submarine lower member on the Main Complex sequence, whose top is frequently seen at elevations up to 100 m asl (very obvious in the east coast of São Nicolau) also provides us with an insight regarding the position of the mean sea-level for its age interval. According to the eustatic curve [27], the mean sea-level for the period between 4.7 and 2.6 Ma [25] is ~10 m below the present level, with a maximum highstand at ~22 m asl (at 3.86 Ma). We therefore infer a minimum positive vertical displacement of 80–110 m for this sequence.

In agreement with this, the extensive lava delta of Terra Chã, probably extruded between 1.7 and 1.1 Ma [25], exhibits a regular passage zone at ~40 m asl, 80 m higher than the mean sea-level for the same interval, and 30 m higher than the maximum highstand at 1.45 Ma [27]. This suggests a recent uplift of at least 30–80 m. The uplift is confirmed by the existence of a well-developed island-rimming raised polygenetic abrasion surface that is frequently covered with younger volcanic products, alluvial fan deposits, and rare marine carbonates.

5.4.8 Santa Luzia and Islets

There are no numerical ages for any of the Santa Luzia formations. The presence of a (Quaternary?) lava delta whose passage zone is close to the present sea-level, and terraces up to 26 m asl, reveal positions within the sea-level range for the Quaternary (see Miller et al. [27]), suggesting that significant uplift is unlikely to have happened.

5.4.9 *São Vicente*

In São Vicente, the almost complete lack of submarine lithologies above the present sea-level suggests that whatever vertical movements affected this island did not result in a positive vertical displacement. The existence of a small number of Quaternary terraces up to elevations of 18 m is not enough to infer recent uplift, since these deposits are within the range of the highstands for the last 0.3 Ma [27], their inferred maximum age.

5.4.10 *Santo Antão*

Similar to São Vicente, the only reliable tracers of sea-level in Santo Antão are of probable Quaternary age and do not go beyond ~ 12 m asl. If we consider the age of the Young Tarrafal volcanic unit (0.3–0.4 Ma), the sediments and pillow-lavas of Ponta do Atum may have recorded the MIS9 or MIS11 sea-level height at ~ 12 m asl [27], suggesting the absence of vertical movements since then.

5.5 Discussion and Conclusions

The Cape Verdes exhibit a considerable number of tracers of relative sea-level positions. These are of sedimentary (mostly shallow marine deposits and coastal terraces) or volcanic origin (submarine volcanic units and lava deltas). Older island edifices with a razed morphology have lost the majority of their older tracers to erosion but have a remarkable record of Quaternary tracers (e.g. Sal, Boa Vista and Maio). More prominent early post-erosional edifices, on the other hand, provide better long term records (e.g. Santiago and São Nicolau).

What the ubiquitous Ancient Complexes represent is still poorly understood. They probably comprise several volcanic episodes grouped under the same unit that are spatially, compositionally and temporally distinct. The advanced state of alteration and erosion and the dense dyke swarms that intrudes those units is an obstacle for deciphering the processes that formed their individual instances. The possible existence in the sequence of extensive portions of uplifted sea-floor, in Maio and Santiago, implies mechanisms of upheaval and/or entrainment of deeply seated blocks upward (of 4,000–6,000 m at least) that are still poorly understood. Understanding these processes, though, is just one aspect of the cryptic mechanism of uplift in island evolution studies. The dominance or abundance of submarine volcanic lithologies in the Ancient Complex of many of the islands suggests that those units correspond to the last stages of the seamount phase, or the initial stages of the emergent island, and were affected by some degree of uplift. The existence of an important unconformity between the Ancient Complexes and the overlying

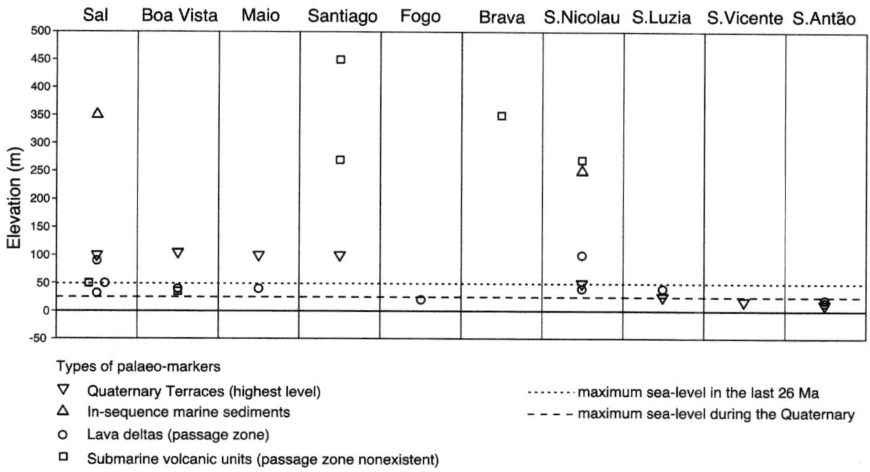

Fig. 5.17 Inferred relative sea-level heights for each island indicating type of palaeo-markers from which they were derived. Maximum sea-level heights for the Quaternary (0–1.81 Ma) and for the last 26 Ma are also shown for comparison [27]. Only Sal, Santiago, Brava and São Nicolau require relative sea-level heights well above the eustatic maximums, denoting significant positive vertical displacements

units—an unconformity that might be associated with subaerial erosion in some cases—suggest volcanic quiescence and exposure to erosion during the shoaling/emersion of the island edifice.

Differential vertical movements can be inferred by the island's geological record (see Figs. 5.17 and 5.18), with two end-members: (1) islands with no significant uplift, like Santo Antão, São Vicente and Santa Luzia; (2) and islands with significant uplift, like Santiago, São Nicolau and Brava, that exhibit positive vertical displacements reaching 450, 280 and 350 m respectively. Santiago and São Nicolau seem to exhibit a history of continuous uplift occurring alongside volcanism. Sal seems to exhibit a complex history with significant early uplift, followed by episodes of lesser uplift and subsidence. The evolutionary histories of Maio and (especially) Boa Vista are still poorly constrained, but they seem to exhibit both uplift and subsidence since their formation. None of the islands seems to have suffered significant post-shield subsidence.

Quaternary marine terraces arranged staircase-like and up to elevations of ~ 105 m asl are ubiquitous on the Cape Verde Islands. They are usually represented by shallow water carbonates, beach deposits (sometimes very extensive), but also as boulder beaches or marine abrasion surfaces. The existence of these deposits suggests a possible recent vertical movement of the islands and can be potentially used to track the recent island vertical motion. The staircase arrangement and their elevations (generally above the Quaternary highstands) suggest a relative regressive system driven by uplift movements. The fact that these features occur at slightly different elevations in different islands, and the fact that some of

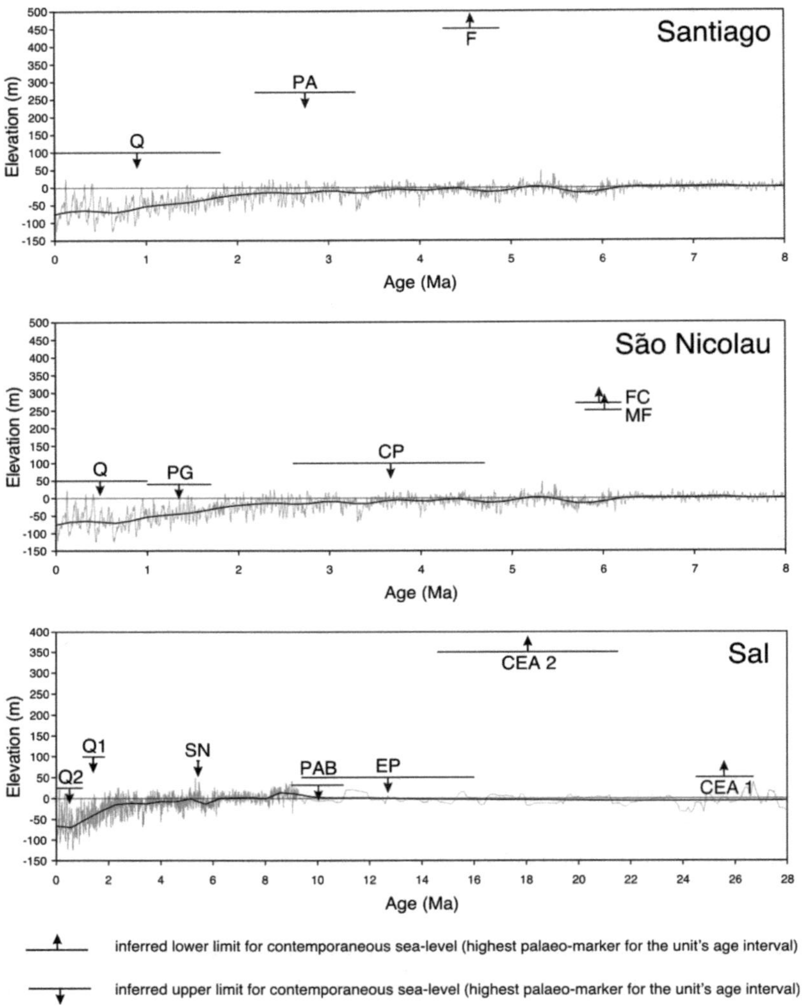

Fig. 5.18 Sea-level height limits for the main geological units of Santiago, São Nicolau and Sal. Only units that have recorded past sea-level are shown. The limits were inferred from the highest palaeo-markers of each unit and age ranges were extracted from the literature (the start of the Quaternary was assumed to be 1.81 Ma). Eustatic curve after Miller et al. [27]. Note that for Santiago and São Nicolau the vertical displacement between the inferred sea-level height and the sea-level from the eustatic curve increases with age, denoting uplift. Sal exhibits a complex history of vertical movements as suggested by the irregular position (with increasing age) of its palaeo-markers. Santiago: F—Flamengos Fm, PA—Pico da Antónia Complex, Q—Quaternary terraces; São Nicolau: MF—Monte Focinho Late Miocene Sediments, FC—Figueira de Coxe Fm, CP—Main Eruptive Complex, PG—Preguiça Fm, Q—Quaternary terraces; Sal: CEA—Ancient Eruptive Complex (1- submarine lavas, 2- carbonates), EP—Principal Eruptive Fm, PAB—Ponta do Altar-Baleia Complex, SN—Serra Negra Fm, Q1—Quaternary terraces prior to Monte Grande-Pedra Lume eruptions, Q2—Quaternary terraces posterior to Monte Grande-Pedra Lume eruptions

the higher levels do not occur in all the older edifices, seems to suggest differences in the uplift histories. This widespread recent uplift trend remains largely unquantified due to the lack of geochronological data.

The amount of uplift we infer (up to 500 m) using the island's exposed geology is similar to the amount of regional uplift proposed by Ali et al. [6] using seismic reflection profiles and flexural calculations on the Cape Verde's moat and volcanic apron. However, our reconstructions suggest a differential and locally restricted uplift rather than a broad regional uplift. Notwithstanding, the geometry of the Cape Verde's moat infill is characterised by a sequence generally tilted away from the islands, rather than towards the islands, suggesting that an upward force partially counteracted island loading and whose origins might be associated with subsurface loading restricted to the island's vicinities (and possibly swell dynamics) [6]. Thus, differential uplift can eventually be partially explained by differential subsurface loading: different amounts of local underplating or intrusions at the base of each edifice can generate uneven upward buoyancy conditions that result in differential uplift. In fact, the existence of differential crustal thickening was inferred by Lodge and Helffrich [58] using receiver function analysis, supporting the possibility of subsurface loading restricted to the vicinities of each island edifice. However, a direct relationship between the amount of crustal thickening inferred by Lodge and Helffrich [58] and the amount of uplift we infer is not immediately evident (e.g. the crust underlying Santo Antão and São Nicolau exhibit similar thickening but the islands exhibit very different uplift trends).

The study of Ali et al. [6] suggests that although surface loading has not caused the expected amount of flexure, the plate is still inflected downwards by $\sim$1700 m and partially compensating the uplift caused by the hotspot swell. Since we do not find evidence for significant subsidence in the islands' exposed sequence we thus infer that subsidence by flexural loading must have occurred in the early stages of edifice building, before the seamounts have reached the ocean's surface.

The more widespread Quaternary uplift might be explained by a broader mechanism affecting the region like thermal buoyancy or the vertical flux associated with hotspot swell dynamics. Further research is necessary to better constrain the archipelago's history of vertical motions and to establish what the role of each mechanism is. However, it is clear that the causes of uplift are not archipelago-wide throughout the Cape Verde's history.

References

1. Parsons, B., & Sclater, J. (1977). An analysis of the variation of ocean floor bathymetry and heat flow with age. *Journal of Geophysical Research (Solid Earth),82*(5), 803–827.
2. Menard, H. (1973). Depth anomalies and the bobbing motion of drifting islands. *Journal of Geophysical Research (Solid Earth),78*(B3), 5128–5137.
3. McNutt, M., & Menard, H. W. (1978). Lithospheric flexure and uplifted Atolls. *Journal of Geophysical Research (Solid Earth),83*(B3), 1206–1212.

4. Scott, G., & Rotondo, G. (1983). A model to explain the differences between Pacific plate island-atoll types. *Coral Reefs,1*(3), 139–150.
5. Schmidt, R., & Schmincke, H.-U. (2000). Seamounts and island building. In H. Sigurdsson, B. Houghton, S. McNutt, H. Rymer, & J. Stix (Eds.), *Encyclopedia of volcanoes* (pp. 383–402). Salt Lake City, UT: Academic Press.
6. Ali, M. Y., Watts, A. B., & Hill, I. (2003). A seismic reflection profile study of lithospheric flexure in the vicinity of the Cape Verde Islands. *Journal of Geophysical Research (Solid Earth),108*(B5), 2239–2263.
7. Rees, B., Detrick, R., & Coakley, B. (1993). Seismic stratigraphy of the Hawaiian flexural moat. *Bulletin of the Geological Society of America, 105*(2), 189–205.
8. Filmer, P., McNutt, M., Webb, H., & Dixon, D. (1994). Volcanism and archipelagic aprons in the Marquesas and Hawaiian Islands. *Marine Geophysical Researches, 16*(5), 385–406.
9. Schmincke, H.-U. (2004). *Volcanism* (1st ed.). Berlin: Springer.
10. Detrick, R., & Crough, T. (1978). Island subsidence, hot spots, and lithospheric thinning. *Journal of Geophysical Research (Solid Earth), 83*(B3), 1236–1244.
11. McNutt, M. (1988). Thermal and mechanical properties of the Cape Verde Rise. *Journal of Geophysical Research (Solid Earth), 93*(B4), 2784–2794.
12. Sleep, N. (1990). Hotspots and mantle plumes: Some phenomenology. *Journal of Geophysical Research (Solid Earth), 95*(B5), 6715–6736.
13. Morgan, J., Morgan, W., & Price, E. (1995). Hotspot melting generates both hotspot volcanism and a hotspot swell. *Journal of Geophysical Research (Solid Earth), 100*(B5), 8045–8062.
14. Burke, K., & Wilson, J. (1972). Is the African Plate stationary? *Nature, 239*(5372), 387–390.
15. Morgan, W. J. (1983). Hotspot tracks and the early rifting of the Atlantic. *Tectonophysics, 94*(1–4), 123–139.
16. Gripp, A., & Gordon, R. (2002). Young tracks of hotspots and current plate velocities. *Geophysical Journal International, 150*(2), 321–361.
17. Crough, S. (1982). Geoid height anomalies over the Cape Verde Rise. *Marine Geophysical Researches, 5*(3), 263–271.
18. Amante, C., & Eakins, B. W. (2009). ETOPO1 1 arc-minute global relief model: Procedures, data sources and analysis. *NOAA Technical Memorandum NESDIS NGDC-24,* 19.
19. Torres, P., Silva, L., Serralheiro, A., Mendes, M., Macedo, J., & Gomes, A. (2002). Geologia da Ilha do Sal. Comunicações do Instituto de Investigação Científica Tropical.
20. Mitchell, J., Bas, M. L., Zielonka, J., & Furnes, H. (1983). On dating the magmatism of Maio, Cape Verde Islands. *Earth and Planetary Science Letters, 64*(1), 61–76.
21. Holm, P., Grandvuinet, T., Friis, J., Wilson, J. R., Barker, A. K., & Plesner, S. (2008). An 40Ar-39Ar study of the Cape Verde hot spot: Temporal evolution in a semistationary plate environment. *Journal of Geophysical Research (Solid Earth), 113*(B8), B08201.
22. Serralheiro, A. (1976). A Geologia da Ilha de Santiago (Cabo Verde). *Boletim do Museu e Laboratorio Mineralógico e Geológico da Faculdade de Ciências, 14*, 157–369.
23. Macedo, J., Serralheiro, A., & Silva, L. (1988). Notícia Explicativa da Carta Geológica da Ilha de S Nicolau (Cabo Verde) na escala de 1:50000, Garcia de Orta. *Serviços Geológicos, 11*(1–2), 1–32.
24. Silva, L., Serralheiro, A., Macedo, J., Gomes, A., & Torres, P. (1990). Carta Geológica de Cabo Verde, Ilha do Sal, na escala de 1/25000 (folhas 1-2), Edição do Instituto Investigação Científica Tropical/Instituto de Cooperação Económica.
25. Duprat, H., Friis, J., Holm, P., Grandvuinet, T., & Sørensen, R. (2007). The volcanic and geochemical development of São Nicolau, Cape Verde Islands: Constraints from field and 40Ar/39Ar evidence. *Journal of Volcanology and Geothermal Research, 162*(1–2):1–19.
26. Plesner, S., Holm, P. M., & Wilson, J. R. (2002). 40Ar-39Ar geochronology of Santo Antão, Cape Verde Islands. *Journal of Volcanology and Geothermal Research, 120*(1–2), 103–121.
27. Miller, K., Kominz, M., Browning, J., Wright, J., Mountain, G., Katz, M., et al. (2005). The Phanerozoic record of global sea-level change. *Science, 310*(5752), 1293–1298.

28. Smellie, J. (2000). Subglacial eruptions. In H. Sigurdsson, B. Houghton, S. McNutt, H. Rymer, & J. Stix (Eds.), *Encyclopedia of volcanoes* (pp. 403–418). Salt Lake City, UT: Academic Press.
29. Trenhaile, A. (2000). Modeling the development of wave-cut shore platforms. *Marine Geology, 166*(1), 163–178.
30. Trenhaile, A. (2002). Rock coasts, with particular emphasis on shore platforms. *Geomorphology, 48*(1–3), 7–22.
31. Torres, P., Silva, L., Serralheiro, A., Tassinari, C., & Munhá, J. (2002). Enquadramento geocronológico pelo método K/Ar das principais sequências vulcano-estratigráficas da Ilha do Sal—Cabo Verde, Garcia de Orta. *Serviços Geológicos, 18*(1–2), 9–13.
32. Ubaldo, M., Silva, L., & Torres, P. (1991). Contribuição geológica e micropaleontologica para o conhecimento do "Complexo Eruptivo Antigo" da ilha do Sal, Arquipélago de Cabo Verde, Garcia de Orta. *Serviços Geológicos, 14*(1–2), 9–14.
33. Berggren, W., Kent, D., & Aubry, M.-P. (1995). A revised cenozoic geochronology and chronostratigraphy. In F. Gradstein, F. Agterberg, J. Ogg, J. Hardenbol, P. Van Veen, J. Thierry, et al. (Eds.) Time scales and global stratigraphie correlation (vol. 54, pp. 129–212). Society of Economic Paleontologists and Mineralogists Special Publication.
34. Ogg, J., Ogg, G., & Gradstein, F. (2008). The concise geologic time scale. Cambridge, UK: Cambridge University Press
35. De Rijk, S., Troelstra, S., & Rohling, E. (1999). Benthic foraminiferal distribution in the Mediterranean Sea. *Journal of Foraminiferal Research, 29*(2), 93–103.
36. Zazo, C., Goy, J., Dabrio, C., Soler, V., Hillaire-Marcel, C., Ghaleb, B., et al. (2007). Quaternary marine terraces on Sal Island (Cape Verde archipelago). *Quaternary Science Reviews, 26*(7–8), 876–893.
37. Bebiano, J. (1932). A geologia do Arquipélago de Cabo Verde. *Comunicações dos Serviços Geológicos de Portugal,18*, 167–187.
38. Lecointre, G. (1963). Sur les terrains sedimentaires de l'île du Sal, Garcia de Orta. *Serviços Geológicos, 11*(2), 275–289.
39. Serralheiro, A. (1968). Formações Sedimentares do Arquipélago de Cabo Verde. Junta de Investigações do Ultramar.
40. Serralheiro, A., Alves, C., Macedo, J., & Silva, L. (1974). Note preliminaire sur la géologie de l'île de Boa Vista (Cap-Vert). *Garcia de Orta. Serviços Geológicos, 1*(3), 53–60.
41. Serralheiro, A. (1970). Geologia da Ilha de Maio (Cabo Verde). Junta de Investigações do Ultramar.
42. Paepe, P. D., Klerkx, J., Hertogen, J., & Plinke, P. (1974). Oceanic tholeiites on the Cape Verde Islands: Petrochemical and geochemical evidence. *Earth and Planetary Science Letters, 22*(4), 347–354.
43. Stillman, C., Furnes, H., Le Bas, M., Robertson, A., & Zielonka, J. (1982). The geological history of Maio, Cape Verde Islands. *Journal of the Geological Society, 139*(3), 347–361.
44. Gerlach, D., Cliff, R., Davies, G., Norry, M., & Hodgson, N. (1988). Magma sources of the Cape Verde Archipelago: Isotopic and trace element constraints. *Geochimica et Cosmochimica Acta,52*, 2979–2992.
45. Davies, G., Norry, M., Gerlach, D., & Cliff, R. (1989). A combined chemical and Pb-Sr-Nd isotope study of the Azores and Cape Verde hot spots; the geodynamic implications. *Geological Society of London Special Publications, 42*(1), 231–255.
46. Martins, S., Mata, J., Munhá, J., Madeira, J., & Moreira, M. (2008). Evidências geológicas e geoquímicas para a existência de duas unidades estratigráficas distintas na Formação do Pico da Antónia (Ilha de Santiago, República de Cabo Verde), Memórias e Notícias. *Universidade de Coimbra, 3*, 123–128.
47. Silva, L., Le Bas, M., & Robertson, A. (1981). An oceanic carbonatite volcano on Santiago, Cape Verde Islands. *Nature, 294*(5842), 644–645.
48. Madeira, J., Mata, J., & Mourão, C. (2006). Volcano-tectonic structure of Brava Island (Cape Verde). In *Livro de Resumos do VII Congresso Nacional de Geologia* (vol. I, pp. 279–282). Portugal: Univ. Évora.

49. Mourão, C., Mata, J., Doucelance, J., Madeira, J., Brum da Silveira, A., Silva, L., & Moreira, M. (2010). Quaternary extrusive calciocarbonatite volcanism on Brava Island (Cape Verde). *Journal of African Earth Sciences, 56*(2–3), 59–74.
50. Serralheiro, A., & Ubaldo, M. (1979). Estudo estratigráfico dos sedimentos do Campo da Preguiça ilha de S Nicolau (Cabo Verde). *Garcia de Orta. Serviços Geológicos, 3*(1–2), 75–82.
51. Bernoulli, D., Hottinger, L., Spezzaferri, S., & Stille, P. (2007). Miocene shallow-water limestones from São Nicolau (Cabo Verde): Caribbean-type benthic fauna and time constraints for volcanism. *Swiss Journal of Geosciences, 100*(2), 215–225.
52. Torres, A., & Soares, J. (1946). Formações Sedimentares do Arquipélago de Cabo Verde I—Actualização de conhecimentos. Junta das Missões Geográficas e de Investigações Coloniais.
53. Bosence, D. (1983). Coralline algal reef frameworks. *Journal of the Geological Society, 140*(3), 365–376.
54. Webster, J., Clague, D., Braga, J., Spalding, H., Renema, W., Kelley, C., et al. (2006). Drowned coralline algal dominated deposits off Lanai, Hawaii; carbonate accretion and vertical tectonics over the last 30 ka. *Marine Geology, 225*(1–4), 223–246.
55. Serralheiro, A. (1966). Contribuição para o conhecimento geológico da Ilha de S. Vicente (Cabo Verde). *Garcia de Orta, Serviços Geológicos, 14*(1), 139–152.
56. Jørgensen, J., & Holm, P. (2002). Temporal variation and carbonatite contamination in primitive ocean island volcanics from São Vicente, Cape Verde Islands. *Chemical Geology, 192*(3–4), 249–267.
57. Holm, P., Wilson, J. R., Christensen, B., Hansen, L., Hansen, S., Khein, K. M., et al. (2006). Sampling the Cape Verde plume: Evolution of melt compositions on Santo Antão, Cape Verde Islands. *Journal of Petrology, 47*(1), 145–189.
58. Lodge, A., & Helffrich, G. (2006). Depleted swell root beneath the Cape Verde Islands. *Geology, 34*(6), 449–452.
59. Bernard-Griffiths, J., Cantagrel, J.-M., Alves, C., Mendes, F., Serralheiro, A., & Macedo, J. (1975). Geochronologie: Donnés radiometriques potassium-argon sur quelques formations magmatiques des îlles de l'archipel du Cap Vert CR. *Seances Academy of Science Series D, 280*, 2429–2432.

Chapter 6
Dating of Sea-Level Palaeo-Markers

6.1 Introduction

In Chap. 4, the main types of palaeo-markers for sea-level were introduced and how they can be used to infer the relative sea-level. The concept of vertical displacement was also introduced: the difference in elevation between the relative sea-level inferred from a particular marker and the contemporaneous eustatic sea-level. Hence, an essential factor associated with the process of estimating the vertical displacement is the accurate determination of the age of the deposit. Later, in Chap. 5, a detailed analysis of the volcanostratigraphy of each island was performed and the main palaeo-markers of sea-level were reported. However, due to the lack of a precise geochronological framework, our uplift reconstructions were qualitative and associated with large uncertainties. Thus, in order to achieve better vertical movement reconstructions, we have collected a number of samples for numerical dating, representative of the main palaeo-markers of sea-level described in Chap. 5. The resulting age results, as well as information about the methods, are presented in this chapter.

6.1.1 Types of Samples for Dating Purposes

The classification system proposed in Chap. 4 mostly concerns the nature and distinctive features of the most common sea-level palaeo-markers found on ocean islands. This classification emerged from the need to systematically describe those features. However, for dating purposes, we propose to group our samples in three main types according to the most suitable methods used in their dating: in-sequence fossiliferous marine sediments, Quaternary fossiliferous marine terraces, and submarine volcanic units. The proposed division arises from the intrinsic limiting factors associated with each dating method (e.g. applicable age range, isotopic system used, etc). The available rock dating methods and their limitations (a more detailed description of each method is presented later) are listed below.

R. A. S. Ramalho, *Building the Cape Verde Islands*, Springer Theses, 139
DOI: 10.1007/978-3-642-19103-9_6,

- Strontium isotope stratigraphy (SIS). This method is suitable for dating marine carbonates and other sediments containing marine fossils constituted by carbonated hard skeletal remains [1–4]. The method is potentially applicable to the entire Phanerozoic, but the numerical control of this technique is best within the 0–40 Ma age interval because during this period the change in $^{87}Sr/^{86}Sr$ with time is very fast and close to being linear, when viewed at a resolution of 5 Ma [5]. However, taking to account the present precision achieved by mass spectrometers, SIS ages lying in the interval between 0 and ~1 Ma have an associated uncertainty that may equal the value of the age. Thus, SIS is, in fact, more appropriate to date samples lying in the 1–40 Ma range.
- U-Th Disequilibrium geochronology. This method is suitable for dating marine carbonates in general but is especially useful to date shallow water coralline structures [6]. In principle, this method can be used to date samples with an age range between 300,000 and 600 000 years [6]. Thus, the application of this method is practically limited to the dating of some of the youngest Quaternary formations.
- $^{40}Ar/^{39}Ar$ geochronology. This method is suitable for dating volcanic samples within the entire Phanerozoic and beyond [7].

In summary, we have three available dating methods: two for marine biogenic carbonates, but with different age ranges, and one for volcanic samples.

Consider the nature and probable age range of the palaeo-markers of sea-level in Cape Verde. It is reasonable to assume that most of the marine sediments intercalated in the volcanostratigraphic sequence of the islands will have an age between the Oligocene/Miocene and the Quaternary. Furthermore, the majority of these sediments will probably lie in the even more restricted interval represented by the Late Miocene and the Plio-Pleistocene, since seven of the ten Cape Verdean islands are probably younger than 10–12 Ma. Hence, the in-sequence sediments (through their fossils) are potentially datable with SIS and lie within the 1–40 Ma interval.

By definition, the Quaternary marine terraces found on the archipelago will lie in an age interval between 0 and ~1.8 Ma. This means that U-Th does not cover their entire age range and SIS does not have the necessary relative resolution to date samples lying in the lower part of the interval. For this reason, both methods should be used in these samples. However, it is highly unlikely that the highest marine terraces (above, for example, 50 m asl) have ages younger than 0.6 Ma because that would imply very high uplift rates and complex uplift/subsidence histories. Thus, for samples collected in terraces above 50 m asl, SIS was the only method used.

For volcanic samples, the choice was clear since we are restricted to a single method that covers the entire age range the samples might exhibit. For these samples $^{40}Ar/^{39}Ar$ geochronology was used.

A selection of the best samples from each group was subsequently dated using the appropriate method. All samples from the in-sequence fossiliferous marine sediments and the Quaternary fossiliferous marine terraces were analysed palaeontologically by collaborators at the London Natural History Museum.

Likewise, a few samples were scrutinized for microfossil analysis by collaborators Dr. Daniela Schmidt (Bristol) and Dr. Haydon Bailey (Network Stratigraphic Consulting). All other sediments that did not contain fossils, such as some of the raised boulder beach deposits associated with marine erosive features, were not dated. Likewise, we did not attempt to date wave-cut surfaces despite the fact that some of them might be dateable with cosmogenic exposure methods.

6.2 Palaeontological Dating of Marine Fossiliferous Sediments

All fossiliferous samples were scrutinised by collaborators at the London Natural History Museum for fossil identification and, if possible, to constrain age and palaeo-water depths. Drs Jon Todd (mollusks), Paul Taylor (bryozoan) and Kenneth Johnson (corals) kindly donated their time for this study. At this stage, the study is not yet concluded. However, a preliminary study soon revealed that any biostratigraphic ages yielded by the collected macrofauna are too broad to be used in uplift reconstructions. Furthermore, several factors contribute to the complexity of this study: (1) many species that are nowadays extinct at higher latitudes (thus providing good age constraints in those places) are extant in present tropical latitudes; (2) because the insular condition favours endemism, many species are restricted to the Cape Verdes and their effective age range is poorly constrained; (3) in Cape Verde the mixing of Atlantic and Mediterranean faunas may occur, making facies identifications more complex. Chapter 5, however, reports on a biostratigraphic success.

6.3 Strontium Isotope Stratigraphy on Carbonates

6.3.1 Introduction

Strontium isotope stratigraphy (SIS) is nowadays widely accepted as a potential way of dating marine sediments [1–3]. This method is based on the fact that seawater strontium isotopic composition has changed with time through the Phanerozoic and we now have a reasonable record of this change through the global standard strontium curve (GSSC) [1–5, 8–10]. The principles behind this method will be briefly described.

Strontium (Sr) is an alkaline earth metal extremely reactive with oxygen and water. It is the 15th most abundant element on earth, averaging 0.034% of all igneous rocks. Due to its similar atomic radius, strontium frequently and readily substitutes for calcium in many minerals. Strontium has four stable, naturally occurring isotopes: ^{84}Sr (0.56%), ^{86}Sr (9.86%), ^{87}Sr (7.0%) and ^{88}Sr (82.58%). From these only ^{87}Sr is radiogenic (but stable) and is generated by decay of ^{87}Rb with a half life of 4.99×10^{10} a. The abundance of ^{87}Sr is expressed, by convention, relative to the abundance of stable ^{86}Sr, as the ratio $^{87}Sr/^{86}Sr$ [1].

The Sr isotopic composition of seawater is greatly influenced by the distinct isotopic ratios in the various source materials, chiefly: the continental crust and the upper mantle [1, 8]. The early evolution of these two geochemical reservoirs is the cause of their distinct isotopic compositions, with higher $^{87}Sr/^{86}Sr$ ratios generated in the continental crust through time as a result of the preferential fractionation of the incompatible elements (thus including Rb) into its mineralogy [1]. The oceans receive inputs from both reservoirs, through different mechanisms. Chemical weathering and riverine flow bring into the oceans Sr derived from the continental crust, while hydrothermal activity along mid ocean ridges and the weathering of basalts provide the main influx derived from the upper mantle [1]. Since the oceans bear a low Rb/Sr ratio, because ^{87}Rb has a long half life, the oceanic residence time of Sr ($\sim 4 \times 10^6$ a) is much longer than the oceanic mixing time ($\sim 10^3$ a), seawater $^{87}Sr/^{86}Sr$ effectively reflects the contribution of the source materials [1]. Thus, the variations of $^{87}Sr/^{86}Sr$ in sea water are a balance between the Sr input from rivers ("continental flux") vs the Sr input from the oceanic hydrothermal systems ("mantle flux") and reflect the tectonic evolution of our planet [11]. If seawater gradually changes in isotopic composition through time, inevitably the sediments formed in ocean basins will entrap that isotopic composition. This is particularly true for carbonates where Sr easily replaces Ca in its mineral structure. As a consequence the ocean's Sr isotopic evolution is documented in the stratigraphic record, allowing the reconstruction of this evolution through the GSSC. In addition, if sediments become a geochemical reservoir for Sr, it also means that sediment recycling also plays an important role in modifying the Sr isotopic composition of seawater. The consequences are that, through sediment recycling in general and particularly through carbonate recycling, the system receives a new input with important effects that need to be taken into account [1]. However, despite the fact that the main contributors to the resulting Sr isotopic composition of seawater are now known, the exact causes for much of the $^{87}Sr/^{86}Sr$ trend during the Phanerozoic are still poorly constrained [4, 5, 9, 12].

The GSSC for the Phanerozoic is now well known, mostly due to the great number of isotopic analyses performed over the last 30 years and the advancements in mass spectrometry (measurements in $^{87}Sr/^{86}Sr$ are now possible to a precision of ± 0.000030 or less). By coupling biostratigraphy, magnetostratigrapy and direct dating (like Ar/Ar) with $^{87}Sr/^{86}Sr$ analyses, the GSSC has achieved a numerical time resolution that continues to improve. The variation in seawater ($^{87}Sr/^{86}Sr$) during the Phanerozoic is characterized by two major trends: a slow decay in $^{87}Sr/^{86}Sr$ with high amplitude fluctuations between the Early Palaeozoic and the Middle Mesozoic, and a rapid and steady rise in $^{87}Sr/^{86}Sr$ from the Middle Mesozoic to the present (see Fig. 6.1). This last trend is of particular importance for dating purposes within the Cenozoic due to the steepness of the curve and its quasi-linearity when viewed at a resolution of 5 Ma [5] (see Fig. 6.2). During the

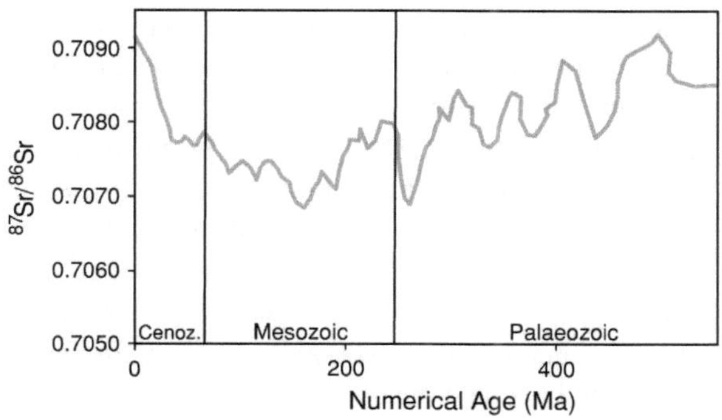

Fig. 6.1 GSSC ($^{87}Sr/^{86}Sr$) for the Phanerozoic, after [4]

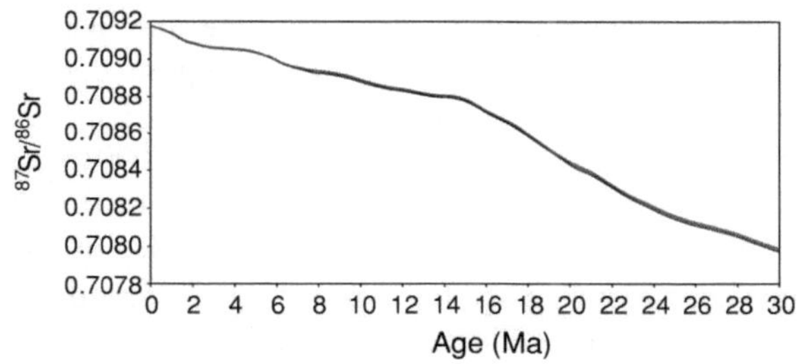

Fig. 6.2 LOWESS GSSC ($^{87}Sr/^{86}Sr$) for the last 30 Ma [4, 5]. ±95% confidence interval shown

Phanerozoic $^{87}Sr/^{86}Sr$ values ranged between 0.7068 (at ~155 Ma) and today's high of 0.709175 [4].

The underlying assumption behind strontium isotope stratigraphy is that the $^{87}Sr/^{86}Sr$ of fossil shells represents the isotopic composition of seawater at the time of their calcification [13]. Calcium carbonate is ideal for this purpose because it contains insignificant amounts of ^{87}Rb so contamination from ^{87}Sr originating from the in situ by decay of ^{87}Rb since carbonate formation is negligible [1, 8]. Thus, if the diagenetic process did not alter the Sr isotopic composition, by determining the $^{87}Sr/^{86}Sr$ of the shell one can match it against the GSSC and infer a correspondent numerical age.

6.3.2 Converting Isotope Ratios into Numerical Ages

In order to simplify the process of converting Sr isotopic ratios into numerical ages, Howarth and McArthur [9] compiled $^{87}Sr/^{86}Sr$ data and fitted to them a nonparametric LOWESS (LOcally WEighted Scatterplot Smoother) statistical regression function [4, 5]. These authors were able to create a GSSC using a variety of published data and their uncertainty: the advantage of the LOWESS regression is that initial weights can be assigned to the dataset in accordance with their analytical errors and accuracy with respect to the NIST SRM 987 standard. Hence, the resulting GSSC gives the best estimate of the mean value of $^{87}Sr/^{86}Sr$ which corresponds to a given numeric age [9]. A lookup table was also provided by Howarth and McArthur [9] from the interpolation of the inverted LOWESS-derived curves as a function of regularly-spaced $^{87}Sr/^{86}Sr$ values. This lookup table, with its 95% confidence intervals, is based on interpolated age values for $^{87}Sr/^{86}Sr$ increments of 0.000001. Later, McArthur et al. [4] further improved the method and updated the data base. For the current work, LOWESS V.4: 08/03 was used. It was kindly provided by Dr. John McArthur upon request.

The process of conversion from $^{87}Sr/^{86}Sr$ ration into numerical ages using LOWESS is a relatively simple one. The process for parts of the curve where its behavior approximates linearity, like the period 0–40 Ma (see Fig. 6.2) is explained in the following lines. Suppose that a particular sample yielded a $^{87}Sr/^{86}Sr$ ratio of 0.709015 with a 2σ uncertainty of ±0.000010. This means that the mean isotopic ratio corresponds to 0.709015 but the confidence interval provided by its uncertainty range is bracketed by 0.709005 (ratio -2σ) and 0.709025 (ratio $+2\sigma$). Using the lookup table provided by McArthur et al. [4], we find that the corresponding age for a mean $^{87}Sr/^{86}Sr$ ratio of 0.709015 is 5.60 Ma. The low and high ratio confidence limits must also be converted into a numerical age, in order to infer the uncertainty of the mean age. This is done by looking up the upper confidence limit column for the age that corresponds to a ratio equal or approximate to the lower sample ratio (ratio -2σ), and look in the lower confidence limit column for the age that corresponds to a ratio equal or approximate to the higher sample ratio (ratio $+2\sigma$). These correspond, respectively, to 5.85 and 5.25 Ma. The resulting numerical age thus corresponds to 5.60 Ma with an asymmetric uncertainty of +0.25 and -0.35 Ma. Figure 6.3 shows graphically how numerical ages are inferred from the LOWESS GSSC [4].

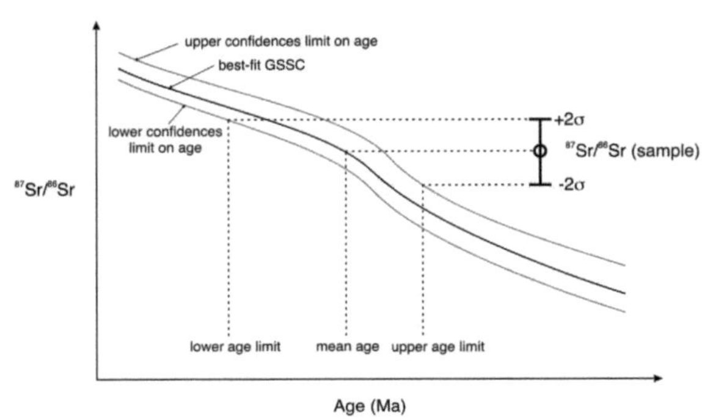

Fig. 6.3 Converting $^{87}Sr/^{86}Sr$ ratios in numerical ages using LOWESS. After McArthur et al. [4]. Note how the upper and lower confidence limits are obtained with lower and higher ratios defined by the 2σ interval

6.3.3 Single Fossil vs Whole Rock Analysis

We opted for single fossil analysis, instead of whole rock analysis, because in whole rock analyses the Sr isotopic composition tends to be contaminated by extraneous phases [14]. Furthermore, the porosity and permeability of a rock, taken as a whole, is frequently higher than that of a single fossil specimen within that rock. This makes the rock more prone to groundwater percolation and supergenic alteration, thus being more liable to isotopic alteration. This is specially true in arid or semi-arid environments (like the Cape Verde Islands) where the formation of calcrete crusts is frequent within any cracks or voids, having a significant influence on the carbonate isotopic composition.

6.3.4 Sample Information

Over 40 samples were chosen for SIS age determination. Tables 6.1 and 6.2 give information about the location, stratigraphic position, elevation and inferred palaeo-water depth of each sample analyzed with SIS geochronology. The abundance of marine carbonates within the volcanostratigraphy of São Nicolau is reflected in the number of samples from this island.

6.3.5 Sample Preparation

Bulk samples were visually checked for suitable macrofossils for dating. Preference was given to bivalves or gastropods with thick shells, exhibiting the lowest degree of alteration/recrystalisation. No preference was given to species or genus utilised for dating.

The macrofossils selected were separated from sediment when enveloped by incoherent matrix or physically extracted from consolidated sediment using a chisel or a micro-drill. The visually best-preserved specimens were picked with the help of a binocular microscope to select clean fragments with minimum surface deposits, diagenetic overgrowth or altered areas. Samples with surface deposits were subsequently mechanically cleaned using a mini rotary tool with a clean grinder bit. Big samples were broken into smaller pieces using a pestle and mortar. All samples were subsequently subjected to an ultrasonic bath for 5 min, washed in ultra pure water and dried for 24 h in a clean environment. A final piece, weighting 0.02–0.04 g, was then chosen for analysis from each sample. These were chemically leached in a 1 M hydrochloric acid solution bath for 30 s in order to remove any possible remaining external contaminants. The samples were again washed in an ultra pure water ultrasonic bath for 5 min, and finally dried.

Table 6.1 Summary of sample information for SIS. Island of São Nicolau

Sample ID	Coordinates UTM WGS84	General location	Dating type
SN01	26 Q 0789244 1835504	Campo da Preguiça	In-sequence marine sediments
SN02	26 Q 0789433 1835259	Campo da Preguiça	In-sequence marine sediments
SN05	26 Q 0815935 1835773	Ribeira Castilhiano	In-sequence marine sediments
SN07	26 Q 0816264 1835836	Ribeira Castilhiano	In-sequence marine sediments
SN10	26 Q 0816368 1834748	Ribeira Antonio Barbacante	In-sequence marine sediments
SN12a	26 Q 0811299 1832747	Ribeira do Palhal (Carriçal)	In-sequence marine sediments
SN12c	26 Q 0811299 1832747	Ribeira do Palhal (Carriçal)	In-sequence marine sediments
SN12d	26 Q 0811299 1832747	Ribeira do Palhal (Carriçal)	In-sequence marine sediments
SN12e	26 Q 0811299 1832747	Ribeira do Palhal (Carriçal)	In-sequence marine sediments
SN13	26 Q 0810114 1832552	Ponta Trancoso	Quaternary terrace
SN14	26 Q 0810093 1832553	Ponta Trancoso	Quaternary terrace
SN15	26 Q 0809931 1832439	Ponta Trancoso	Quaternary terrace
SN16	26 Q 0809694 1832452	Ponta Trancoso	Quaternary terrace
SN17	26 Q 0809415 1832681	Ponta Trancoso	Quaternary terrace
SN22	26 Q 0816346 1832692	Talhada	In-sequence marine sediments
SN26	26 Q 0795548 1841461	Coelha	Quaternary terrace
SN27	26 Q 0795527 1841647	Coelha	Quaternary terrace
SN28	26 Q 0794919 1841616	Coelha	Quaternary terrace
SN30a	26 Q 0789609 1835002	Campo da Preguiça	In-sequence marine sediments
SN31	26 Q 0789739 1831860	Turil	In-sequence marine sediments
SN33	26 Q 0789691 1831391	Ribeira da Ponta da Pataca	In-sequence marine sediments
SN35	26 Q 0789776 1831457	Ribeira da Ponta da Pataca	In-sequence marine sediments
SN38	26 Q 0790798 1833507	Preguiça	In-sequence marine sediments

Samples with a lower case letter after the reference (e.g. SN30a) correspond to different levels in the same sedimentary package

(Continued)

Table 6.1 (Continued)

Stratigraphic position	Elevation (m)	2σ (m)	Sedimentary type	Palaeo-depth (m)
Within Late Miocene marine Sed.	223	5	Transition/offshore?	5–20
Within Late Miocene marine Sed.	200	5	Transition/offshore	20–50
Base of main eruptive complex	83	3	Coralline algal reef	15–30
Base of Preguiça Formation	21	2	Transition/offshore	20
Base of main eruptive complex	77	5	Coralline algal reef	15–30
Base of Preguiça Formation	14	1	Foreshore/shoreface	3–5
Base of Preguiça Formation	10	1	Foreshore/shoreface	5–9
Base of Preguiça Formation	18	1	Foreshore/shoreface	1
Base of Preguiça Formation	14	1	Foreshore/shoreface	3–5
Quaternary sediments	26	2	High energy pebbly beach	0–10
Quaternary sediments	26	2	High energy pebbly beach	0–10
Quaternary sediments	29	2	High energy pebbly beach	0–10
Quaternary sediments	25	2	High energy pebbly beach	0–10
Quaternary sediments	30	2	Foreshore/shoreface?	0–5
Within main eruptive complex	80	3	Incipient coral reef	0–15
Quaternary sediments	34	2	High energy pebbly beach	0–10
Quaternary sediments	29	2	High energy pebbly beach	0–10
Quaternary sediments	21	2	High energy pebbly beach	0–10
Within Late Miocene marine Sed.	163	5	Transition/offshore	60–100
Within Main Eruptive Complex	22	2	Transition/offshore	50–70
Within Main Eruptive Complex	25	2	Transition/offshore	50–70
Within Main Eruptive Complex	22	2	Transition/offshore	50–70
Base of Preguiça Formation	8	1	Foreshore/shoreface	0–5

For stratigraphic units refer to Chap. 5. Palaeo-depth constraints presented in this table are based on the criteria defined in Chap. 4 and local field evidence

6.3.6 Sample Chemistry

After the cleaning process, the samples were dissolved in 3 M HNO_3 in order to oxidise organic matter. We extracted 10% of the sample for posterior trace element analysis to check the degree of alteration of the analysed sample. The remaining sample solution was centrifuged and the solid residue was discarded. The samples were then dried on a 120°C hot plate. The dry residue was then converted to chloride salt by repeatedly dissolving it in 6 M HCl and drying on a hot plate. The dry residue was dissolved in 1 M HCl and purified by a cation column chromatography using a standard procedure to separate the main cations from the solution. Samples were subsequently dried again and converted to a nitric salt. The next step was to dissolve

Table 6.2 Summary of sample information for SIS. Other islands

Sample ID	Coordinates UTM WGS84	General location	Dating type
ST03	27 p 0204809 1691609	Ribeira do Fontão (Tarrafal)	In-sequence marine sediments
ST05	27 p 0205014 1691597	Ribeira do Fontão (Tarrafal)	In-sequence marine sediments
ST07	27 p 0205288 1691977	Ribeira do Fontão (Tarrafal)	In-sequence marine sediments
ST08	27 p 0205867 1692056	Ribeira do Fontão (Tarrafal)	In-sequence marine sediments
ST10	27 p 0236044 1663893	Senhora da Luz	Quaternary terrrace
ST16	27 p 0209154 1696227	Ponta Furna	In-sequence marine sediments
ST18	27 p 0205797 1697077	Ponta Moreia	In-sequence marine sediments
ST23	27 p 0233159 1649605	Ponta das Bicudas	In-sequence marine sediments
BV01	27 Q 0295446 1771748	Trigonometric point Condinho	Quaternary terrace
BV18	27 Q 0297872 1783367	Rabil	Quaternary terrace
MO01	27 p 0264573 1675252	Casa Pad	Quaternary terrace
MO03	27 p 0267381 1674521	Barreiros	Quaternary terrace
MO06	27 p 0262610 1675931	Montinho do Lume	Quaternary terrace
MO07	27 p 0263855 1679127	Palmeira	Quaternary terrace
SA04	26 Q 0679662 1874326	Ponta do Atum	In-sequence marine sediments
SL02	27 Q 0298157 1842688	Ribeira da Fragata	In-sequence marine sediments
SL08	27 Q 0294406 1842063	Lomba Branca	Quaternary terrace
SL12	27 Q 0287572 1858246	Buracona	In-sequence marine sediments
SV01	26 Q 0707314 1861731	Monte Passarinho	Quaternary terrace
SV03	26 Q 0724849 1860204	Ribeira de Curral de J. Paula	Quaternary terrace

Sample ID reflects island of origin
ST Santiago, *BV* Boa Vista, *MO* Maio, *SA* Santo Antão, *SL* Sal, *SV* São Vicente

(Continued)

Table 6.2 (Continued)

Stratigraphic position	Elevation (m)	2σ (m)	Sedimentary type	Palaeo-depth (m)
Within Pico da Antónia E. C.	21	3	Transition/offshore	50–70
Within Pico da Antónia E. C.	36	3	Transition/offshore	50–70
Within Pico da Antónia E. C.	38	3	Transition/offshore	50–70
Within Pico da Antónia E. C.	69	3	Transition/offshore	50–70
Quaternary sediments	18	1	High energy pebbly beach	0–10
Base or within Pico da Antónia E. C.	148	5	Transition/offshore	50–70
Base or within Pico da Antónia E. C.	128	5	Transition/offshore	50–70
Within Pico da Antónia E. C.	8	1	Incipient coral reef	5–15
Pleistocene sediments	25	3	Foreshore/shoreface	0–5
Pleistocene sediments	38	3	Foreshore/shoreface	0–5
Quaternary sediments	48	3	Foreshore/shoreface	0–5
Quaternary sediments	44	3	Foreshore/shoreface	0–5
Quaternary sediments	25	3	Foreshore/shoreface	0–5
Quaternary sediments	77	5	Incipient coral reef	5–15
Within Yount Tarrafal Volcanics	19	1	Transition/offshore	0–15
Ribeira de Fragata Formation	48	3	Transition/offshore	0–10
Quaternary sediments	40	3	Foreshore/shoreface	0–5
Within Ponta do Altar-Baleia E. C.	7	1	Foreshore/shoreface	0–5
Quaternary sediments	18	1	High energy pebbly beach	0–10
Quaternary sediments	12	1	High energy pebbly beach	0–10

For stratigraphic units refer to Chap. 5. Palaeo-depth constraints presented in this table are based on the criteria defined in Chap. 4 and local field evidence

the samples in 3 M HNO_3 and to process them using Sr-spec chromatography. This last step was taken in order to remove, as best as possible, the residual Rb and Ca in solution and to end up with a pure Sr solution for mass spectrometric analysis.

6.3.7 Mass Spectrometry

Mass spectrometric analyses were performed on the ThermoFinnigan Triton thermal-ionisation multicollector mass spectrometer (TIMS) at the Bristol Isotope Group laboratory. This mass spectrometer is equipped with nine movable collectors. Sr compositions were measured in static mode. A total of 297 cycles of 4.2 s each were used per sample measurement. Raw $^{87}Sr/^{86}Sr$ ratios were corrected for ^{87}Rb interference on ^{87}Sr—always negligible due to the low Rb content of carbonate and the efficient chemical separation. Mass spectrometric mass discrimination was corrected by normalizing to $^{87}Sr/^{86}Sr = 0.1194$ using an exponential law.

Samples were diluted in 10% HNO_3 solution and were loaded on a degassed standard purity Re ribbon filament, previously prepared with $TaCl_5$ and H_3PO_4, and desiccated. Standards were loaded in the same way, using a NIST SRM 987 solution. Sample filaments were immediately loaded in the magazine and the magazine was inserted in the vacuum chamber and pumping initiated and left overnight to ensure proper vacuum conditions. A minimum of five standards for 21 total positions was used. The sequence was initiated with two standards and then a standard was intercalated between every five samples. A total of 19 analyses of NIST SRM 987, over the period of 6 months, gave $^{87}Sr/^{86}Sr = 0.710246 \pm 0.000022(2\sigma)$. All uncertainties reported for samples incorporate only internal errors.

Major and trace element analysis were also performed to assess sample quality. For these analyses we used a Finnigan Element 2 high-resolution single collector mass spectrometer. Based on repeat analyses of consistency standards, trace element/Ca ratios are considered precise to approximately 5%.

6.3.8 Quality Control

A few indicators have been traditionally used to assess diagenetic alteration. The use of dull cathodoluminescense has been considered as a possible way to distinguish between pristine and recrystallised samples [15]. Another indicator, perhaps the most widely used, concerns the concentration of Mn and Fe in the biogenic carbonate. These parameters are based on the principle that the recrystallisation of carbonate normally occurs via diagenetic reactions that result in changes, generally an increase, in the concentrations of Mn and Fe [16–19]. According to McArthur et al. [16], there is no real standard threshold in element concentrations with which diagenetic alteration can be assessed, but some values can be used as guidelines. These authors suggest using the following approximate concentrations as guidelines for upper limits beyond which diagenetic alteration may be unacceptable: 300 ppm for Mn and 500 ppm for Fe, in calcite, and 400 ppm of Mg in aragonite (concentrations of Mg in calcite have a great range and cannot be used as a diagnostic for diagenetic alteration). Another parameter that has been suggested as a possible way to evaluate diagenetic alteration is the Sr/Ca ratio of the sample, specifically, a check if this is the expected value for a pristine marine carbonate (somewhere between 1 and 1.5 mmol/mol for calcite) [9, 20]. However, considerable variability in Sr/Ca ratios may occur in present day calcitic bivalve shells due to kinetic processes, suggesting that this parameter should be used with care [21]. Moreover, it has been reported that shells that the guidelines deemed pristine sometimes exhibit Sr isotopic signatures that do not reflect the contemporaneous composition of seawater, so that no obvious correlation is found between the departure of $^{87}Sr/^{86}Sr$ from expected values and the degree of alteration [15, 16, 22]. Notwithstanding these caveats, in order to assess the possible alteration, samples were analysed for several trace

Table 6.3 Main trace element composition of analysed samples. Samples from São Nicolau

Sample ID	Mn/Ca	Mn (ppm)	Fe/Ca	Fe (ppm)	Sr/Ca	Sr (ppm)
SN01	**3,586**	**1,972**	**8,033**	**4,498**	1.0	901
SN02	114	62	55	31	1.0	887
SN05	**652**	**359**	**1,494**	**837**	0.61	534
SN05(2)	**1,033**	**568**	**3,523**	**1,973**	0.30	263
SN07	67	37	58	33	0.78	685
SN07(3)	23	12	6.6	3.7	1.2	1,038
SN10	**1,132**	**623**	505	283	0.40	353
SN12a	0.97	0.53	11	6.3	1.6	1,394
SN12c	4.2	2.3	9.7	5.4	1.1	988
SN12d	4.7	2.6	25	14	1.9	1,675
SN12e	10	5.6	32	18	0.35	309
SN13	0.48	0.26	3.7	2.1	2.3	2,002
SN14	1.4	0.75	8.9	5.0	1.6	1,382
SN15	0.82	0.45	7.9	4.4	2.3	2,002
SN15(2)	1.9	1.1	1.5	0.84	2.2	1,950
SN16	0.55	0.30	0.21	0.12	0.60	530
SN17	4.4	2.4	3.3	1.9	1.2	1,045
SN19	2.8	1.5	1.0	0.56	1.8	1,600
SN22	5.3	2.9	13	7.2	0.57	501
SN22(2)	2.7	1.5	34	19	0.25	217
SN26	0.13	0.07	−3.9	−2.2	0.65	568
SN27	0.36	0.20	0.79	0.44	0.63	552
SN28	0.26	0.15	0.13	0.07	0.60	526
SN30	146	81	344	193	0.95	834
SN31	4.7	2.6	34	19	0.41	364
SN33	126	69	43	24	0.44	385
SN33(2)	26	14	68	38	0.42	366
SN35	6.2	3.4	38	21	0.72	637
SN38	5.0	2.7	16	8.9	1.1	969

Samples with a lower case letter after the reference (e.g. ST18a) correspond to different levels in the same sedimentary package; samples with numbers in parenthesis (e.g. ST07(2)) correspond to analyses on different fossils from the same stratigraphic level. Bold entries correspond to values above suggested guidelines for sample quality

elements including Mn, Fe, Sr and Mg. Since the majority of the organisms we analysed are calcite formers, we assume that Mg is less important for the purpose.

The results for Mn, Fe and Sr are summarised in Tables 6.3 and 6.4. Using the concentration of Mn and Fe as possible indicators of diagenetic alteration, only five samples seem to have concentrations above the suggested values: SN01, SN05, SN05(2), SN10 and SL04. However, if one uses the Sr/Ca as an indicator of diagenetic alteration, many samples lie outside the interval suggested as corresponding to pristine material.

In addition, a few different fossils from the same stratigraphic levels were also analysed and compared to check reproducibility. Likewise, different organisms were analysed in search for less altered ratios.

Table 6.4 Main trace element composition of analysed samples. Samples from Santiago (ST), Boa Vista (BV), Maio (MO), Santo Antão (SA), Sal (SL) and São Vicente (SV)

Sample ID	Mn/Ca	Mn (ppm)	Fe/Ca	Fe (ppm)	Sr/Ca	Sr (ppm)
ST03	8.69	4.8	32	18	0.61	541
ST03(2)	28	15	28	16	0.93	815
ST05	4.7	2.6	12	6.9	0.69	603
ST07	81	45	137	77	0.43	381
ST07(2)	126	70	79	44	0.45	398
ST08	138	76	60	34	0.37	329
ST10	248	136	142	80	0.45	398
ST16	3.0	1.7	46	26	0.53	464
ST16(2)	16	8.6	32	18	0.59	518
ST18a	6.4	3.5	44	25	0.95	832
ST18a(2)	5.5	3.0	40	22	1.1	963
ST18b	13	7.2	215	120	0.30	264
ST23a	5.4	2.9	105	59	1.9	1,694
ST23b	21	11	362	203	1.1	1,002
BV01	3.1	1.7	9.7	5.5	0.75	656
BV18	2.9	1.6	11	6.2	0.44	387
MO01	11	5.9	76	42	0.94	830
MO01(2)	12	6.5	23	13	1.1	958
MO03	4.8	2.6	4.0	2.2	0.80	707
MO03(2)	20	11	22	12	0.68	599
MO03(3)	1.6	0.89	4.5	2.5	0.61	535
MO06	0.10	0.06	1.1	0.62	1.3	1,129
MO07	1.6	0.87	−1.9	−1.1	2.7	2,359
SA04	4.5	2.4	77	43	1.2	1,089
SA04(2)	3.0	1.6	47	27	1.3	1,137
SL02	**652**	**359**	357	200	0.24	208
SL08	103	57	194	109	0.52	457
SL12	0.69	0.38	−1.8	−1.0	1.5	1,354
SV01	0.30	0.16	2.9	1.6	1.5	1,291
SV03	0.69	0.38	18	10	0.61	532

Samples with a lower case letter after the reference (e.g. ST18a) correspond to different levels in the same sedimentary package; samples with numbers in parenthesis (e.g. ST07(2)) correspond to analyses in different fossils from the same stratigraphic level. Bold entries correspond to values above suggested guidelines for sample quality

6.3.9 Results and Discussion

All sample $^{87}Sr/^{86}Sr$ ratios are presented in Tables 6.5 and 6.6, with uncertainties and information about the analyzed material. The typical analytical error associated with the reported analyses is 0.000010–0.000011. All $^{87}Sr/^{86}Sr$ ratios were converted to numerical ages by using LOWESS v4: 08/03 lookup table [4, 9].

Table 6.5 Sample $^{87}Sr/^{86}Sr$ ratios and correspondent SIS ages with age-converted uncertainties. Samples from São Nicolau (SN)

Sample ID	Analyzed material	$^{87}Sr/^{86}Sr$	±uncertainty (2σ)	age (Ma)	+unc	-unc
SN01	Bivalve	0.707873	0.000011	32.85	0.3	0.35
SN02	Bivalve	0.705878	0.000010	–		
SN05	Bivalve/gastropod	0.708475	0.000011	19.48	0.24	0.22
SN05(2)	Echinoderm	0.707930	0.000011	31.50	0.6	0.5
SN07	Bivalve	0.709047	0.000010	4.54	0.51	1.34
SN07(3)	Bivalve	0.708946	0.000011	7.33	1.07	0.54
SN10	Rhodolith	0.707269	0.000011	>40		
SN12a	Bivalve/gastropod	0.709021	0.000011	5.46	0.3	0.42
SN12c	Bivalve/gastropod	0.709057	0.000010	3.52	1.16	1
SN12d	Bivalve/gastropod	0.709016	0.000010	5.59	0.25	0.36
SN12e	Echinoderm	0.707563	0.000010	>40		
SN13	Bivalve/gastropod	0.709113	0.000011	1.364	0.245	0.185
SN14	Bivalve/gastropod	0.708851	0.000011	11.18	0.96	0.63
SN15	Bivalve	0.708680	0.000011	16.68	0.31	0.31
SN15(2)	Bivalve	0.708591	0.000011	17.98	0.22	0.21
SN16	Bivalve	0.709092	0.000010	1.768	0.442	0.27
SN17	Bivalve	0.708950	0.000010	7.18	0.75	0.49
SN19	Bivalve/gastropod	0.709103	0.000011	1.534	0.318	0.221
SN22	Bivalve/gastropod	0.708816	0.000010	12.93	1.42	0.73
SN22(2)	Bryozoan	0.707715	0.000010	>40		
SN26	Oyster	0.709153	0.000011	0.668	0.306	0.326
SN27	Oyster	0.709127	0.000011	1.175	0.188	0.218
SN28	Oyster	0.709164	0.000010	0.382	0.308	0.372
SN30	Bivalve	0.708985	0.000011	6.13	0.32	0.21
SN31	Bivalve	0.709003	0.000010	5.84	0.21	0.25
SN33	Bivalve	0.708881	0.000011	10.15	0.56	0.5
SN33(2)	Pectinid	0.708874	0.000011	10.37	0.58	0.52
SN35	Bivalve	0.709019	0.000010	5.52	0.26	0.38
SN38	Bivalve	0.709086	0.000011	1.97	0.45	0.378

Samples with a lower case letter after the reference (e.g. SN30a) correspond to different levels in the same sedimentary package; samples with numbers in parenthesis (e.g. SN22(2)) correspond to analyses on different fossils from the same stratigraphic level

Converted ages and respective uncertainties are also presented in Tables 6.5 and 6.6. The resulting numerical ages range from >40 Ma to ~0.4 Ma.

A closer look at the results immediately reveals that ages, in general, contrast greatly with the reported ages for the archipelago's volcanism. Furthermore, the SIS dates are incompatible with other, better-established stratigraphic and geochronological constraints. For comparison, Tables 6.7 and 6.8 present the yielded SIS ages with the possible and probable ages of each formation, inferred from stratigraphic relations and the geochronological data from the literature. Since the majority of the in-sequence carbonates of Cape Verde occur within or between volcanostratigraphic units that are well mapped, whose general

Table 6.6 Sample $^{87}Sr/^{86}Sr$ ratios and correspondent SIS ages with age-converted uncertainties. Samples from Santiago (ST), Boa Vista (BV), Maio (MO), Santo Antão (SA), Sal (SL) and São Vicente (SV)

Sample ID	Analyzed material	$^{87}Sr/^{86}Sr$	±uncertainty (2σ)	Age (Ma)	+unc	−unc
ST03	Bivalve	0.708760	0.000011	15.37	0.27	0.3
ST03(2)	Bivalve	0.708959	0.000010	6.84	0.56	0.41
ST05	Oyster	0.709077	0.000011	2.27	0.53	0.472
ST07	Bivalve	0.708844	0.000011	11.52	0.93	0.73
ST07(2)	Bivalve	0.708710	0.000010	16.16	0.28	0.26
ST08	Bivalve/gastropod	0.708605	0.000011	17.81	0.22	0.23
ST10	Bivalve	0.708911	0.000011	9.25	0.54	0.88
ST16	Bivalve	0.708623	0.000011	17.59	0.22	0.26
ST16(2)	Bivalve	0.708672	0.000011	16.82	0.31	0.31
ST18a	Bivalve	0.709039	0.000011	4.89	0.44	0.85
ST18a(2)	Bivalve	0.708304	0.000011	22.19	0.27	0.27
ST18b	Rhodolith	0.709019	0.000010	5.52	0.26	0.38
ST23a	Bivalve	0.708901	0.000010	9.58	0.47	0.63
ST23b	Rhodolith	0.708993	0.000012	6.01	0.26	0.26
BV01	Bivalve/gastropod	0.709115	0.000011	1.337	0.232	0.187
BV18	Bivalve/gastropod	0.709064	0.000013	2.78	1.69	0.6
MO01	Bivalve/gastropod	0.708966	0.000011	6.59	0.57	0.34
MO01(2)	Bivalve/gastropod	0.709107	0.000010	1.461	0.256	0.186
MO03	Bivalve/gastropod	0.709087	0.000011	1.93	0.45	0.356
MO03(2)	Bivalve/gastropod	0.708827	0.000011	12.45	0.79	0.95
MO03(3)	Bivalve/gastropod	0.708890	0.000011	9.89	0.54	0.53
MO06	Bivalve/gastropod	0.709114	0.000010	1.35	0.219	0.171
MO07	Bivalve/gastropod	0.709136	0.000011	1.04	0.202	0.292
SA04	Bivalve	0.708252	0.000011	23.03	0.37	0.32
SA04(2)	Bivalve	0.708756	0.000010	15.43	0.25	0.26
SL02	Bivalve/gastropod	0.708285	0.000011	22.47	0.31	0.27
SL08	Bivalve/gastropod	0.708968	0.000011	6.54	0.53	0.33
SL12	Bivalve/gastropod	0.709089	0.000011	1.86	0.46	0.325
SV01	Bivalve	0.709080	0.000009	2.18	0.38	0.409
SV03	Bivalve/gastropod	0.709089	0.000010	1.86	0.43	0.305

Samples with a lower case letter after the reference (e.g. ST18a) correspond to different levels in the same sedimentary package; samples with numbers in parenthesis (e.g. ST07(2)) correspond to analyses on different fossils from the same stratigraphic level

chronology is known, we can estimate the possible age range within which these features were formed. In some cases, due to stratigraphic proximity to levels dated with other methods (mostly K/Ar and Ar/Ar), we can even narrow this interval a bit more and estimate a probable age interval. On rarer occasions we have independent age data for the same formations, through previous biostratigraphic analysis (from the literature). Likewise, for a few Quaternary terraces, we can compare the SIS ages with the our own U-Th geochronology data. Below, a few examples are explored in detail.

Table 6.7 Comparison between SIS ages and expected ages from other stratigraphic constraints present in the literature and provided by this study

Sample ID	SIS age (Ma)	+unc	−unc	Possible age range[a]	Probable age[b]
SN01	32.85	0.3	0.35	**11.8–5.8**	**6.2–5.8**
SN02	–			**11.8–5.8**	**6.2–5.8**
SN05	19.48	0.24	0.22	5.8–5.0	5.8–5.0
SN05(2)	31.5	0.6	0.5	5.8–5.0	5.8–5.0
SN07	4.54	0.51	1.34	2.7–1.0	1.7–1.0
SN07(3)	7.33	1.07	0.54	2.7–1.0	1.7–1.0
SN10	>40			5.8–5.0	5.8–5.0
SN12a	5.46	0.3	0.42	2.7–0.78	1.0–0.78
SN12c	3.52	1.16	1	2.7–0.78	1.0–0.78
SN12d	5.59	0.25	0.36	2.7–0.78	1.0–0.78
SN12e	>40			2.7–0.78	1.0–0.78
SN13	1.364	0.245	0.185	<1.0	<0.78
SN14	11.18	0.96	0.63	<1.0	<0.78
SN15	16.68	0.31	0.31	<1.0	**0.660–0.389**
SN15(2)	17.98	0.22	0.21	<1.0	**0.660–0.389**
SN16	1.768	0.442	0.27	<1.0	**0.258–0.219**
SN17	7.18	0.75	0.49	<1.0	<0.78
SN19	1.534	0.318	0.221	<1.0	<0.78
SN22	12.93	1.42	0.73	5.1–2.7	3.5–3.7
SN22(2)	>40			5.1–2.7	3.5–3.7
SN26	0.668	0.306	0.326	<1.8	<0.6
SN27	1.175	0.188	0.218	<1.8	<0.6
SN28	0.382	0.308	0.372	<1.8	**0.273–0.243**
SN30	6.13	0.32	0.21	**11.8–5.8**	**6.2–5.8**
SN31	5.84	0.21	0.25	5.1–2.7	5.1–3.1
SN33	10.15	0.56	0.5	5.1–2.7	5.1–3.1
SN33(2)	10.37	0.58	0.52	5.1–2.7	5.1–3.1
SN35	5.52	0.26	0.38	5.1–2.7	5.1–3.1
SN38	1.97	0.45	0.378	2.7–0.78	2.7–0.78

[a]Age ranges inferred by the works of Macedo et al. [24], Bernoulli et al. [22], Duprat et al. [25] and this study

[b]Probable ages based on the possible age ranges but narrowed due to the proximity to other age data or because the formation was dated using other methods (in bold)

Consider samples SN01, SN02 and SN30. Sample SN01 exhibits high quantities of Mn and Fe, probably denoting pervasive weathering, but SN02 and SN30 exhibit much lower values (see Tables 6.3 and 6.4). These three samples were collected in different but nearby outcrops of the same sedimentary package: the Late Miocene Marine Sediments in São Nicolau. As the name suggests, this formation was assigned to the Late Miocene on the basis of its micro-fossil content [23, 24]. A recent biostratigraphic study by Bernoulli et al. [22] confirmed the Late Miocene age proposed by Serralheiro and Ubaldo [23] and assigned the 11.8–5.8 Ma interval as the correspondent numerical age for the deposition of these sediments. According to Bernoulli et al. [22] the interval of deposition might even be restricted

Table 6.8 Comparison between SIS ages and expected ages from the stratigraphic constraints present in the literature and provided by this study

Sample ID	SIS age (Ma)	+unc	−unc	Possible age range[a]	Probable age[b]
ST03	15.37	0.27	0.3	3.3–2.2	3.3–2.2
ST03(2)	6.84	0.56	0.41	3.3–2.2	3.3–2.2
ST05	2.27	0.53	0.472	3.3–2.2	3.3–2.2
ST07	11.52	0.93	0.73	3.3–2.2	3.3–2.2
ST07(2)	16.16	0.28	0.26	3.3–2.2	3.3–2.2
ST08	17.81	0.22	0.23	3.3–2.2	3.3–2.2
ST10	9.25	0.54	0.88	<1.8	<1.0
ST16	17.59	0.22	0.26	8.5–2.2	3.3–2.2
ST16(2)	16.82	0.31	0.31	8.5–2.2	3.3–2.2
ST18a	4.89	0.44	0.85	8.5–2.2	3.3–2.2
ST18a(2)	22.19	0.27	0.27	8.5–2.2	3.3–2.2
ST18b	5.52	0.26	0.38	8.5–2.2	3.3–2.2
ST23a	9.58	0.47	0.63	3.3–2.2	2.5–2.2
ST23b	6.01	0.26	0.26	3.3–2.2	2.5–2.2
BV01	1.337	0.232	0.187	<1.8	<1.8
BV18	2.78	1.69	0.6	<1.8	<1.8
MO01	6.59	0.57	0.34	<1.8	<1.8
MO01(2)	1.461	0.256	0.186	<1.8	<1.8
MO03	1.93	0.45	0.356	<1.8	<1.8
MO03(2)	12.45	0.79	0.95	<1.8	<1.8
MO03(3)	9.89	0.54	0.53	<1.8	<1.8
MO06	1.35	0.219	0.171	<1.8	<1.8
MO07	1.04	0.202	0.292	<1.8	<1.8
SA04	23.03	0.37	0.32	0.4–0.2	0.4–0.2
SA04(2)	15.43	0.25	0.26	0.4–0.2	0.4–0.2
SL02	22.47	0.31	0.27	9.0–5.5	6.5–5.5
SL08	6.54	0.53	0.33	<1.8	<1.8
SL12	1.86	0.46	0.325	16–9.0	11–9.0
SV01	2.18	0.38	0.409	<1.8	<1.8
SV03	1.86	0.43	0.305	<0.3	<0.3

[a]Age ranges inferred by the work of Bernard-Griffths et al. [44], Holm et al. [41], Plesner et al. [45], Torres et al. [42] and this study

[b]Probable ages based on the possible age ranges but narrowed due to proximity to other age data

to 6.2–5.8 Ma if future studies confirm the presence of *Globigerinoides conglobatus*. In addition, the sediments are covered by the submarine lavas of Figueira de Coxe Formation. These lavas were emplaced when the sediments were still soft (as attested to by the presence of deformation marks on their top surfaces) and in some places sedimentation was partially synchronous with the volcanic activity. According to Duprat et al. [25] the Figueira de Coxe Formation yields ages between 6.2 and 5.7 Ma, an age in agreement with our own Ar/Ar geochronology. Thus, it is reasonable to assume that the most probable numerical age for the Late Miocene Marine Sediments corresponds to ~6.2 Ma. Yet SN01 yielded a $^{87}Sr/^{86}Sr$ of 0.707873 ± 11, suggesting an age of $32.85 + 0.3/-0.35$ Ma (SN02 yielded an unrealistically lower ratio without any correspondent in the GSSC). This value is surprisingly similar to the $^{87}Sr/^{86}Sr = 0.70817 \pm 7$ ($\sim$32.80 Ma) reported by Bernoulli et al. [22] for the exact same sediments, and a value these authors discarded on the grounds that it conflicts with the biostratigraphic age suggested by the fossil content. Like these authors, we are forced to conclude that the analyzed $^{87}Sr/^{86}Sr$ does not represent the original seawater signature. Curiously, sample SN30 yielded a $^{87}Sr/^{86}Sr = 0.708985 \pm 11$ which corresponds to an age of $6.13 + 0.32/-0.21$ Ma, and a value very similar to the expected age.

Now consider samples SN13, SN14, SN15, SN15(2) and SN16. Their Mn and Fe concentrations are well within the suggested alteration limits and their Sr/Ca is variable. All five samples correspond to Quaternary marine terraces collected within an area of 1 km^2 and at similar elevations. All five samples lie over subaerial lavas attributed to the Preguiça Formation and dated with Ar/Ar (this study) to around 0.78 ± 0.05 Ma. Samples SN15 and SN16 were also dated with U-Th (this study) and yielded ages that correspond, respectively, to $0.496 + 0.195/-0.076$ and 0.283 ± 0.020. However, their SIS ages range between 1.364 and 17.98 Ma. SN14 is an excellent example of a sample that exhibits trace element concentrations well within the suggested parameters (including a Sr/Ca $\sim$1.58) but yields a lower-than-expected $^{87}Sr/^{86}Sr$ incompatible with its stratigraphic position.

In a similar way, samples ST16, ST16(2), ST18a, ST18a(2) and ST18b were collected in different locations within the same sedimentary package. All seem to present trace element concentrations within the suggested limits, yet they exhibit a large variability in the strontium ages (between 4.89 and 22.19 Ma). Furthermore, samples ST18a, ST18a(2) and ST18b were collected within centimeters of each other.

It is also evident that fossils with more open structures like echinoderm spicules, rhodoliths and bryozoans, exhibit the ratios that deviate more from the expected ratios [e.g. samples SN05(2), SN10, SN12e, and SN22(2)]. These samples also yielded the lowest Sr/Ca ratios, perhaps denoting extreme leaching associated with the higher porosities.

In summary, the majority of the analysed samples exhibit $^{87}Sr/^{86}Sr$ ratios that are lower than expected for their stratigraphical position. The deviation between their expected and analyzed ratios is inconsistent and does not seem to correlate with any obvious trace element concentration. Figure 6.4 plots the amount of deviation between the expected age range and the strontium age versus the concentration of Mn, Fe and Sr (relative to Ca), to check if it is possible to establish a correlation between the concentration of these elements and the amount of alteration in the $^{87}Sr/^{86}Sr$ composition. Unfortunately no evident correlation seems to be present, because no relationship of any form arises from the plots.

In order to test internal variability within individual shells, we have analysed selected samples by microprobe. Samples MO07, SN02, SN12c, SN26, SN28, SN30 and ST10 were scanned in this way. Concentrations of Mg, Si, Ca, Mn, Fe, Sr and C (and their oxidised forms) were scanned both across and along the shells' growth layers, in 500–700 μm intervals. However, internal variability was very small,

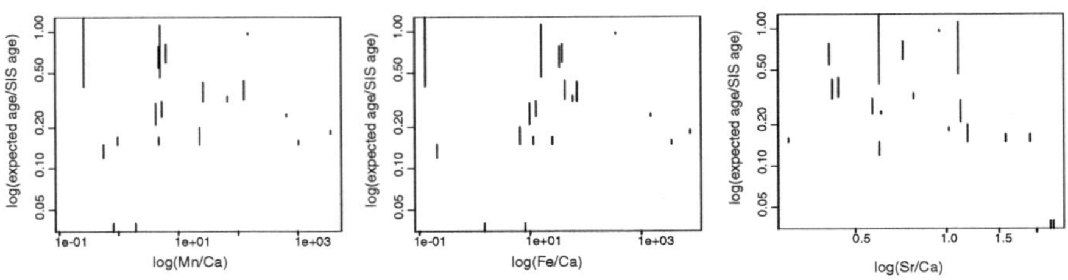

Fig. 6.4 Expected age/SIS age vs Mn, Fe and Sr concentrations. *Bars* correspond to range defined by the oldest expected age/(SIS age $+2\sigma$) and the youngest expected age/(SIS age -2σ). Note that there is no clear relationship between the concentration of Mn, Fe or Sr in a sample, and the difference between the yielded and expected age

perhaps denoting pervasive alteration. Likewise, along the same scanning lines, we measured the $^{87}Sr/^{86}Sr$ ratios with laser ablation in order to test the internal variability of the Sr isotopic composition of shells. The results once again showed little ($<10^{-4}$) variation in the $^{87}Sr/^{86}Sr$ ratios of shells, however.

The alterations in $^{87}Sr/^{86}Sr$ ratios may be several. Elderfield [1] and Hess et al. [8] suggested that Sr may be exchanged between marine carbonates and coexisting clays, older or younger sediments, and basalts through the percolation of fluids. Since diagenetic or authigenic fluids have different $^{87}Sr/^{86}Sr$ ratios than seawater, the Sr isotopic composition of marine carbonates may be altered [1, 8]. This process may be enhanced by the semi-arid conditions of Cape Verde, as attested to by the abundance of calcrete formations found in the vicinities of many of the sampled formations. Another process that may have contributed to the lower $^{87}Sr/^{86}Sr$ ratios found in these samples is the emplacement of submarine lavas. A large portion of the sedimentary formations included in this study occur above and, more importantly, below submarine lavas. It is very plausible that the emplacement of these lavas induced hydrothermal circulation that changed the original Sr isotopic composition of the underlying sediments. Since the $^{87}Sr/^{86}Sr$ composition of basalts typically ranges 0.7035–0.7040 [26, 27], the interaction of hydrothermal fluids driven by the lava emplacement could result in a reduction of the $^{87}Sr/^{86}Sr$ ratios of the sediments. Consistent with this, note that the recorded SIS ages are all shifted to older values. Given the monotonic rise of the seawater curve through the Cenozoic, this finding is completely consistent with the contamination of the carbonates with basaltic Sr.

6.3.10 Summary

A considerable number of carbonates representing in-sequence marine sediments and Quaternary terraces were sampled and dated by Sr isotopic geochronology. The resulting ages, however, are incompatible with the known stratigraphical constraints and contrast greatly with the few ages yielded by other dating methods.

Variability between fossils from the same sampled formation also seems to be great and in excess of the analytical uncertainty associated with the dating method. Moreover, the deviation between the expected $^{87}Sr/^{86}Sr$ ratios and the observed $^{87}Sr/^{86}Sr$ ratios does not seem to correlate with any particular diagenetic alteration indicator suggested in the literature. The inconsistent but ubiquitous lowering of $^{87}Sr/^{86}Sr$ ratios in the analysed samples might be explained by supergenic alteration, as attested to by the frequent presence of calcrete crusts in the vicinities of the sampled sediments, or by hydrothermal alteration driven by the emplacement of submarine lavas above many of the deposits. Thus, the samples' $^{87}Sr/^{86}Sr$ ratios do not reflect the contemporaneous seawater Sr isotopic composition, and ages based on them are unreliable.

6.4 Laser Ablation U/Th Disequilibrium Geochronology on Fossil Corals

6.4.1 Introduction

U/Th dating is a technique finding wider application that involves calculating ages from the radioactive decay and ingrowth relationships among ^{238}U, ^{234}U and ^{230}Th isotopes. In principle, this dating technique may be used to date materials as young as 3 years and in excess of 600,000 years, but with a precision rarely achieved by other dating methods for the same period [6]. Among the materials available to be used for U/Th dating, coralline aragonite proved to be particularly suitable for the purpose. Thus, it is not surprising that U/Th dating of shallow water corals gained a special popularity among oceanographic, palaeoclimatic and uplift studies.

The method is based on the initial portion of the ^{238}U decay chain, in which ^{238}U decays through a series of intermediate daughters to ^{234}U and to ^{230}Th [6]. In decay chains like this, if a system remains closed to isotopic exchange for time scales longer than the half-lives of the intermediate daughters, a state of secular equilibrium is reached in which the activity of all the nuclides of the chain are equal [6, 28]. However, if a natural process disrupts the secular equilibrium through the fractionation of nuclides in the decay chain, the subsequent trend to equilibrium is a function of time [6]. Thus, if the isotope ratios immediately after the fractionation event are known, it is possible to calculate the elapsed time through the equations of radioactive decay and ingrowth [6]. Consequently, the fundamental issue in this process is to know the original isotopic ratios and especially the amount of daughter isotope in the system. It is in the resolution of this issue that corals become important: corals normally exhibit very low concentrations of initial daughter isotopes but incorporate large amounts of U, making then ideal for dating purposes [6].

Table 6.9 Summary of sample information for U/Th. Islands are Santiago (ST), São Nicolau (SN), Santo Antão (SA), São Vicente (SV) and Brava (BR)

Sample ID	Coordinates UTM WGS84	General Location	Elevation (m)	2σ (m)	Palaeo-depth (m)
ST10	27 P 236044 1663893	Senhora da Luz	18	1	0–10
ST13	27 P 234152 1666562	Praia Baixa	5	1	0–10
SN15	26 Q 809931 1832439	Ponta Trancoso	29	2	0–10
SN16	26 Q 809694 1832452	Ponta Trancoso	25	2	0–10
SN28	26 Q 794919 1841616	Coelha	21	2	0–10
SA01	26 Q 679117 1873512	Ponta do Atum	7	1	0–10
SV03	26 Q 724849 1860204	Rib. de curral de J. Paula	12	1	0–10
BR05	26 p 749833 1647672	Furna	3	1	0–10

Laser ablation U/Th geochronological techniques enable rapid, in situ analyses on solid samples [29]. With laser ablation it is also possible to achieve a better spatial resolution than with micro-mill or conventional techniques [29, 30]. In our analyses we used the method described by Hoffmann et al. [30].

6.4.2 Sample Information

A total of eight coral samples from the Quaternary marine terraces of Cape Verde were analysed. The small number of samples available for this dating method is a direct result of the difficulty in finding corals pristine enough for dating. Information about the location, elevation and palaeo-water depth of each sample is summarised in Table 6.9.

6.4.3 Sample Preparation

Coral samples were were cut with a diamond coated wire saw and subsequently polished with ultra-thin sandpaper. Samples were then washed in a MQ ultrasound bath to remove potential surface contamination, and dried in a clean environment. The surfaces were cleaned with methanol. Pieces were visually checked under a binocular microscope and the pieces with less visible pore-filling contamination, or with a more pristine look, were chosen for analysis.

6.4.4 Mass Spectrometry

All MC-ICPMS measurements were performed in the Bristol Isotope Group (BIG) laboratory using a ThermoFinnigan Neptune with a multi ion counter (MIC) array

Table 6.10 U-Th disequilibrium ages for São Nicolau Quaternary terraces

Sample ID	$(^{230}Th/^{238}U)_A$	$(^{234}U/^{238}U)_A$	U-Th age (ka)	$(^{234}U/^{238}U)_{A0}$
ST10[a]	1.029 ± 0.020	1.017 ± 0.008	–	–
ST13[a]	2.484 ± 0.075	1.092 ± 0.016	–	–
SN15	1.049 ± 0.015	1.046 ± 0.005	$465.7 + 195 - 76$	$1.174 + 0.124 - 0.031$
SN16	1.012 ± 0.015	1.072 ± 0.004	283.6 ± 19.7	1.160 ± 0.010
SN28	0.988 ± 0.013	1.070 ± 0.006	258.3 ± 15.1	1.146 ± 0.011
SA01[a]	1.089 ± 0.014	1.065 ± 0.006	617.9 ± 369.7	1.375 ± 0.375
SV03[b]	0.996 ± 0.023	1.030 ± 0.009	343.9 ± 61.7	1.080 ± 0.022
SV03(2)[b]	0.987 ± 0.015	1.026 ± 0.007	335.0 ± 38.2	1.068 ± 0.015
BR05	0.686 ± 0.008	1.092 ± 0.004	106.05 ± 3.8	1.124 ± 0.005
BR05(2)	0.689 ± 0.007	1.105 ± 0.003	104.42 ± 3.7	1.141 ± 0.004
BR05(3)	0.686 ± 0.006	1.100 ± 0.003	104.50 ± 3.8	1.134 ± 0.004

All uncertainties are 1σ

[a]These samples are out of range for U/Th

[b]The initial $^{234}U/^{238}U$ ratios of these samples indicate open system, so ages are not reliable

and coupled with a New Wave Research UP193HE ArF Excimer laser system. The laser system has a wave-length of 193 nm and a typical power density of $5J/cm^2$ at 70% power output. The laser frequency chosen and spot size depended on the U and Th concentration, but typical values correspond to a frequency of 7 Hz and a spot size of 180–250 μm. Samples were loaded in a standard New Wave Research laser ablation (LA) cell, with purpose-made sample holders. One standard position was destined for NIST SRM 610 and NIST SRM 612 glass mounted in epoxy resin. A carbonate standard was used and corresponds to a secular equilibrium calcite 'standard' for correction of instrumental biases of laser ablation U-Th isotope measurements on $CaCO_3$. Potential matrix effects due to differences between aragonite and calcite are negligible within uncertainties achieved for U-Th isotope measurements using the laser ablation technique [30]. After tuning, a typical measurement sequence consisted of background intensity measurements without ablating, followed by measurements on the carbonate standard. Background intensities (without ablating) were found to be negligible. Sample measurements followed. Ablation is done using He as carrier gas which is mixed with Ar sample gas and N_2 in a quartz mixing cell before injection into the Ar plasma. For U-Th isotope LA measurements presented in this study we used 7 Hz repetition rate and a 250 μm spot size. A LA measurement was done on a 0.5 mm long track, ablated by moving the laser spot at a speed of 20 μm/s along the track in six passes. A standard–sample–standard bracketing procedure was applied and data collection and corrections were done according to Hoffmann et al. [30].

6.4.5 Results and Discussion of Results

The isotope ratios and the resulting ages are summarised in Table 6.10. Three samples were found to be out of range of U/Th dating: ST10, ST13 and SA01.

Samples SV03 and SV03(2) exhibit initial $^{234}U/^{238}U$ ratios that denote open system, so ages are not reliable. All initial $^{234}U/^{238}U$ ratios should be close to those of seawater, at 1.149 ± 0.002 [31, 32]. If they are not then post growth addition or loss of U is inferred and the age is not valid due to the resultant resetting of the isotopic clock. Notwithstanding this caveat, both samples show a very small variability around a possible age of 0.335–0.343 Ma. The three Quaternary terraces we sampled in São Nicolau proved to be within the U/Th age range and their ages are comprehended between 0.258 ± 0.030 Ma and $0.466 + 0.390 - 0.152$ Ma. Perhaps surprisingly, samples SN16 and SN28 seem to have been formed during MIS 8, a low stand. Sample SN15 unfortunately presents an uncertainty too high to infer the MIS stage of formation. Samples BR05, BR05(2) and BR05(3) all correspond to the same deposit in Brava and present very limited variability. Thus, the age of this deposit is accurately attributed to MIS 5c at ~0.105 Ma.

6.4.6 Summary

Eight Quaternary marine terraces from different islands were used for U/Th dating. From these, only four yielded reliable ages and three are from the island of São Nicolau. The resulting ages span from ~0.105 to ~0.466 Ma and all the samples increase in age with increasing elevations. From the resulting four ages, three can be assigned to particular marine isotopic stages: BR01 to MIS 5c, SN16 and SN28 to MIS 8.

6.5 Laser Step-Heating $^{40}Ar/^{39}Ar$ Geochronology on Lavas

6.5.1 Introduction

$^{40}Ar/^{39}Ar$ geochronology is one of the most widely applicable and precise methods in the field of isotope geochemistry [7, 33]. It has been successfully applied to materials that range from single minerals, through whole volcanic rocks, even to extra-terrestrial rocks.

Argon is a noble gas whose main terrestrial isotopes include ^{40}Ar (99.6%), ^{39}Ar (0.337%), and ^{38}Ar (0.063%). The stable ^{40}Ar isotope is generated by the decay of the naturally occurring ^{40}K with a half life of 1.25 Ga making this one of the most useful isotopic systems for geochronology [7]. In conventional K/Ar geochronology techniques the initial content of K had to be measured independently through wet chemistry and flame photometry, and then noble gas spectrometry followed to measure the daughter isotopes [34]. However, the development of the $^{40}Ar/^{39}Ar$ method simplified the process by only using noble gas spectrometry. In the $^{40}Ar/^{39}Ar$ method, K-bearing samples are irradiated in a nuclear reactor in order to produce ^{39}Ar from ^{39}K, thus allowing the estimation of the original K content in the samples. Subsequently, the ratio between the naturally occurring ^{40}Ar and the reactor produced ^{39}Ar (used as a proxy for K) is measured and the age can be calculated since this ratio is proportional to the age of the sample [7, 35]. Another great advantage of $^{40}Ar/^{39}Ar$ dating is that argon can be released in increments by stepwise heating of irradiated samples, producing a spectrum of dates related to the thermal history of the rock and allowing the pseudo-physical separation of sample phases [7, 33, 34]. The step-heating and the separation of phases has the advantage that excess and atmospheric argon may be released in different steps than decay-originated argon, thus allowing a better precision in dating. Excess atmospheric argon, incorporated during mineral growth is a common problem in $^{40}Ar/^{39}Ar$ geochronology because it leads to the presence of unsupported ^{40}Ar and erroneously old ages. Excess argon is less common in volcanic systems where outgassing provides a release mechanism [34], but nonetheless exists.

Recently, laser step-heating $^{40}Ar/^{39}Ar$ methodologies have gained importance because they avoid the vacuum breakdown problems exhibited during conventional furnace step heating techniques [34] and because they involve lower background values (Michael Cosca, Personal Communication, 2009). Another advantage of using laser heating lies in the fact that whole rock samples can be analysed without the need for rock disaggregation and grain separation. For this study we use the method developed by our collaborator Dr. Michael Cosca from the USGS (formerly at the University of Lausanne), which consists of whole rock IR laser step-heating $^{40}Ar/^{39}Ar$ analysis, with 50 incremental steps. The use of 50 incremental steps is of great advantage because, if one has to discard any step due to excess argon, there is still a great number of points to build the regression curve. This yields more accurate results than conventional 5–10 step methodologies and higher confidence in the analyzes.

6.5.2 Sample Information

A total of 15 samples were analysed by $^{40}Ar/^{39}Ar$ geochronology. We strategically focused geochronological studies on the islands of Santiago and São Nicolau because these islands exhibit unique volcanostratigraphic records, rich in palaeo-markers of sea-level. Thus, 12 samples were dedicated to date palaeo-markers in Santiago and São Nicolau. The remaining three samples correspond to particular targets in other islands that were originally chosen for reconciliation with the SIS ages. However, since the SIS ages proved unreliable, their use for uplift study is limited. Their results are presented nonetheless.

Sample information is presented in Table 6.11, and a brief description of the sample localities follows (for a map with sample locations refer to Figs. 7.4, 7.5).

6.5.2.1 Santiago

Flamengos Formation. The Flamengos Formation was sampled close to the highest point where outcrops of this unit could be found: in the Portal da Furna

Table 6.11 Summary of sample information for $^{40}Ar/^{39}Ar$ geochronology

Sample ID	Coordinates UTM WGS84	General Location	Type	Stratigraphic Unit	Elevation (m)	2σ (m)	Palaeo--depth (m)
ST14	26 P 209737 1696097	Ponta Furna	s.f.	PAEC	153	5	27
ST17	26 P 209158 1696222	Ponta Furna	s.f.	PAEC	151	5	29
ST19	26 P 205821 1697012	Ponta Moreia	s.f.	PAEC	131	5	49
ST20	26 P 210347 1655934	Porto Mosquito	s.f.	PAEC	76	2	3
ST25	26 P 212120 1675155	Portal da Furna	s.f.	FFm	422	5	23
ST26	26 P 212247 1675348	Portal da Furna	s.f.	FFm	416	5	29
SN06	26 Q 815935 1835773	Castilhiano	s.f.	MEC	84	5	18
SN08	26 Q 816213 1835814	Castilhiano	s.f.	PFm	39	2	1
SN15	26 Q 809931 1832439	Ponta Trancoso	t	Qsed	29	2	5
SN16	26 Q 809694 1832452	Ponta Trancoso	t	Qsed	25	2	5
SN18	26 Q 810325 1832793	Rib. Gombeza	sa.f.	PFm	15	1	0
SN23	26 Q 816340 1832690	Talhada	s.f.	MEC	92	3	10
SN25	26 Q 797674 1840562	Fig. Coxe	s.f.	FCFm	185	5	80
SN28	26 Q 794919 1841616	Coelha	t	Qsed	21	2	5
SN36	27 Q 789776 1831457	Rib. Pta Pataca	s.f.	MEC	23	2	70
SA02	26 Q 679124 1873511	Ponta do Atum	s.f.	YTV	12	1	0
SL05	27 Q 298258 1842750	Serra Negra	sa.f.	SNFm	88	3	0
MO08	27 P 266824 1674788	Rib. da Cumiassa	s.f.	CVFm	37	3	3

s.f. submarine flows, *t* terrace, *sa.f.* subaerial flow, *PAEC* Pico da Antónia Eruptive Complex, *FFM* Flamengos Fm, *MEC* Main Eruptive Complex, *PFm* Preguiça Fm, *Qsed* Quaternary sediments, *FCFm* Figueira de Coxe Fm, *YTV* Young Tarrafal Volcanics, *SNFm* Serra Negra Fm, *CVFm* Casas Velhas Fm

area (Assomada). In this area the submarine volcanics reach elevations of ~450 m asl and are unconformably covered by subaerial lavas corresponding to the Pico da Antónia Eruptive Complex and Assomada/Monte das Vacas formations [36]. The submarine sequence is composed of clusters of individual pillows floating in a weathered hyaloclastitic breccia. Two samples (ST25 and ST26) were collected in two different clusters of pillows at 422 and 416 m asl, in the eastern slope of Portal da Furna col, towards the Flamengos valley. Since the younger subaerial flows fill palaeo-topographic lows in the Flamengos volcanics, it is not possible to quantify how much of this unit was eroded, making it difficult to estimate the palaeo-depth of emplacement. To estimate the palaeo-depth we measured the vertical distance to the unit's local highest point, along its erosional unconformity with overlying units. This distance was 23 and 29 m respectively for ST25 and ST26.

Pico da Antónia Eruptive Complex. We chose two areas to sample Pico da Antónia Eruptive Complex: the northern peninsula, near Tarrafal, and the western coast, near Porto Mosquito.

The plateau that forms the northern peninsula of Santiago comprises from the base to the top [37]: (1) carbonatitic/nephelinitic extrusive complex (Ancient Eruptive Complex), (2) shallow marine epiclastic carbonatitic sediments (Órgãos Formation), (3) bioclastic calcarenites, and (4) Pico da Antónia Eruptive Complex plateau basalts. The basaltic pile is composed of thick submarine sheet flows and pillow-lavas (Pico da Antónia lower member), and is covered by later subaerial flows (Pico da Antónia upper member). The submarine flows infilled the underlying topography and their tops, still visible, form a regular surface at 175–190 m asl. This surface can be used as a marker for the relative sea-level at the time. Three samples of the submarine flows were collected from two locations in Achada Costa (ST14 and ST17) at 153 and 151 m asl, and from Ponta Moreia (ST19) at 131 m asl. Their palaeo-depths were inferred using the vertical distance to the top of the submarine flows at $\sim$180 m asl (27, 29 and 49 m, respectively). These values are compatible with the possible palaeo-depth of the underlying bioclastic sediments (<50 m). The later subaerial lavas flowed on top of this surface and spilled to lower levels, suggesting that sea-level was lower during their extrusion.

The Porto Mosquito area, on the western side of the volcanic edifice, exhibits a sequence of Pico da Antónia Eruptive Complex products (attributed to the upper member). The exposed sequence comprises a lower package composed of pillow-lavas and hyaloclastites dipping seaward and reaching an elevation of $\sim$80 m, and an upper package of subaerial flows. From this area we sampled (ST20) at 76 m, 3 m below the subaerial-submarine interface.

6.5.2.2 São Nicolau

Figueira de Coxe Formation. Sample SN25 was collected in the right bank of the Figueira de Coxe valley, about 300 m to the northeast of Figueira de Coxe's settlement, at an elevation of 185 m asl. In this location the Figueira de Coxe volcanics are characterised by north-dipping hyaloclastite breccias and pillow-lavas. The sequence is found covering Ancient Eruptive Complex outcrops (in the bottom of the valley) and is covered by the Main Eruptive Complex subaerial flows close to the top of the ridge [24]. SN25 was collected in one of the clusters of pillows north of the footpath that follows eastwards from Figueira de Coxe's settlement. Since the top of the Figueira de Coxe Formation is eroded, we used the vertical distance to the highest nearby point where this formation outcrops to estimate the palaeo-depth of the sampled pillows. Just 300 m to the east, the Figueira de Coxe Formation reaches its highest point at 260–260 m asl. Thus we estimated a palaeo-depth of ~75 m for SN25. No higher pillow-lavas were sampled due to access difficulties and the highly weathered condition of the sequence.

Main Eruptive Complex. The passage zone between the subaerial and submarine lavas of the Main Eruptive Complex can be best observed in the eastern part of the island, while in the central and western part of the island this interface is usually covered by young flows [38].

In the easternmost part of São Nicolau, a headland of the Main Eruptive Complex was "fossilized" by the younger and extensive lava delta of Terra Chã (attributed to the Preguiça Formation). However, the Main Eruptive sequence is well exposed on the coast west of Ponta Barroso and in the deep canyons of Ribeira de Castilhiano, Ribeira de António Barbacante, and Ribeira de Covoada de Bodela. The Main Eruptive Complex products comprise submarine lavas in continuity with their subaerial counterparts. The passage zone is clearly visible and can be found very regularly at ~100 m asl in all these localities. We have paid special attention to the sequences in Castilhiano and Talhada, where we collected two samples of the submarine lavas.

In Castilhiano the exposed sequence constitutes, from the base to the top: (1) Ancient Eruptive Complex (?) intensively weathered (submarine?) tuffs outcropping up to elevations of ~80 m asl; (2) marine carbonates interpreted as shallow marine 10–50 m depth lying unconformably over the previous unit, up to 6 m thick [38]; (3) Main Eruptive Complex submarine lavas, composed of a 3 m thick submarine sheet flow immediately overlying the carbonates, with a hyaloclastite and pillow breccia above it, up to an elevation of 102 m asl; and (4) subhorizontal Main Eruptive Complex subaerial flows. One sample (SN06) was collected in the submarine sheet flow immediately over the carbonates at an elevation of 84 m asl. The inferred palaeo-depth for this sample is 18 m, the vertical distance to the local passage zone at 102 m. This estimate of palaeo-depth is compatible with the inferred palaeo-depth of the underlying carbonates.

In Talhada/Ponta Barroso region, the sequence is composed of: Ancient Eruptive Complex (?) up to 20 m asl; marine carbonates up to 5 m thick lying unconformably over the previous unit; Main Eruptive Complex submarine flows from 5–20 m asl up to ~ 100 m asl; and Main Eruptive Complex subaerial flows. Marine calcarenites can also be found close to the top of the submarine sequence. Sample SN23 was collected in the submarine flows at 92 m asl just 10 m below the passage zone and immediately above the calcarenite bed. The inferred palaeo-depth is 10 m.

In the central and western part of the island the Main Eruptive Complex submarine lavas are usually covered by young flows and the passage zone is difficult to observe. We have studied the sequence in the Turil area, where the top of the submarine flows reach ~90 m asl. These outcrops reveal a sequence composed of submarine lavas (pillows and sheet flows) with occasional lenses of marine carbonate beds. Sample SN36 was collected in the submarine lavas of Ribeira da Ponta da Pataca, immediately above the carbonates, at 23 m asl. The vertical distance between these lavas and the top of the submarine sequence is ~70 m. This is the value we assumed as palaeo-depth.

Young Lava Deltas and Quaternary Marine Terraces. In the eastern part of São Nicolau it is possible to find several valley-filling lava sequences that created prominent lava deltas when they reached the coast (interpreted as part of Preguiça Formation). The most outstanding of these structures is the Terra Chã lava delta, that forms the eastern tip of the island [38].

The Terra Chã flows probably originated in a WNW-ESE lineament of vents that includes Monte Caldeirinha and Morro cones. They in-filled the existing east-draining valleys and spilled out to form the volcanic delta, extending the island considerably to the east. The contemporaneous relative sea-level must have been ~40 m asl because the passage zone in the delta sequence is found very regularly at this elevation, separating the eastwards steep-dipping breccia and pillow foresets from the flat lying subaerial flows above. Marine sediments can be found at the base of this sequence, separating it from the underlying units. We have sampled this unit near the Castilhiano river mouth. Sample SN08 was collected in a pillow-lava from the lava delta, located 50 m upstream from the sea and at an elevation of 42 m asl, just 1 m below the local passage zone.

Other prominent lava deltas occur in the Carriçal area, west of Terra Chã. In this region, a wavecut surface was incised in the Main Eruptive sequence and later covered by the younger flows and sediments. The Main Eruptive sequence is composed of seaward dipping foresets of hyaloclatite breccia and pillows with subaerial subhorizontal lavas on top. The passage zone is visible at ~70 m asl. The wavecut surface was carved in the submarine lavas and is currently at 15–30 m asl, extending up to 600 m inland. It is covered by a package of subaerial lava flows that spilled from the valleys of Ribeira do Palhal, Ribeira de Gombeza and Ribeira Sóca; the latter created the distinctive Carriçal effusive fan. A package of sediments can be found separating the two sequences: further inland they comprise a thin layer of stream deposits, while near the Ribeira do Palhal river mouth (near Carriçal) they comprise a >8 m thick pile of estuarine sediments. This suggests that the contemporaneous coastline was not too far from the current coastline and that the sea-level height was slightly above the present level (~15 m). We did not find any submarine morphologies in the young lavas because these were probably excised by modern marine erosion. Sample SN18 corresponds to one of the subaerial flows that covered the older abrasion surface and was collected in the river mouth of Ribeira de Gombeza at 20 m asl.

6.5.2.3 Santo Antão

Sample SA02 was collected in Ponta do Atum circa 500 m SSW of the old tuna factory. The sample was collected in the submarine lavas that cover the fan-delta sedimentary package (see Fig. 5.16), and as near as possible to the passage zone ($\sim$12 m asl). This marker corresponds to the sea-level contemporaneous with the extrusion of the youngest lavas of the Young Tarrafal volcanic unit of Holm et al. [39].

6.5.2.4 Sal

Sample SL08 was collected on the left bank of the Ribeira de Fragata, near its source. The sampled unit is a subaerial lava just 1 meter above the passage zone of the Serra Negra Formation, at 88 m asl. This passage zone accurately marks the contemporaneous sea-level height.

6.5.2.5 Maio

We sampled the submarine lavas of Casas Velhas Formation in the highest point where these crop out. Sample MO08 was collected in Ribeira da Cumiassa in South Maio, circa 2.1 km upstream from the river mouth. The unit is very weathered but preserved clusters of pillow-lavas crop out up to 35–40 m asl, denoting the minimum position for the contemporaneous sea-level.

6.5.3 *Sample Preparation*

All samples were visually checked in thin section for freshness. Any samples exhibiting highly altered olivines were discarded. A 2 mm chip was extracted from the interior of each sample and sent for irradiation.

6.5.4 *Mass Spectrometry*

The $^{40}Ar/^{39}Ar$ analyses were performed at the Institute of Mineralogy and Geochemistry at the University of Lausanne. Approximately 1 mm^3 of washed but otherwise untreated whole rock samples, together with standards, were irradiated for 10 MWH with cadmium lining in the CLICIT facility at the Oregon State University TRIGA reactor. Monitoring of the neutron flux was done using Fish Canyon Tuff sanidine, using an age of 28.20 $\pm$ 0.08 Ma [40]. Isotopic production ratios were determined from irradiated CaF_2 and KCl salts. For this irradiation, the following production values were measured: $^{(36/37)}Ca = 0.0002964 \pm 0.00000587$; $^{(39/37)}Ca = 0.00080 \pm 0.000053$; and $^{(38/39)}K = 0.0122 \pm 0.000028$. The irradiated cubes of basalt and standards were loaded into 3 mm wells within a stainless steel planchette attached to a fully automated extraction line evacuated to UHV conditions. Samples were degassed using a 20W CO_2 laser with a slightly defocused beam. The gas was expanded and purified by exposure to a cold finger maintained at $-132°C$, and a hot SAES GP50 getter. The purified gas was expanded into a Nu Instruments Noblesse mass spectrometer and argon isotopes were measured using a Faraday detector for m/e 40, and ion counting multipliers for m/e 39, 38, 37, and 36. Fifty (50) step cycles were used; for each step cycle, data were collected for a period of 300 s: 180 s for laser step heating and 120 s of repose time. Blanks were measured between samples or between steps if an issue was suspected. Time zero intercepts were determined by best-fit regressions to the data. Detector intercalibration was done using repeated analysis of atmospheric argon. Data and ages reported in Table 7.2 have been corrected for blanks, mass discrimination, detector intercalibration, radioactive decay subsequent to irradiation, and interfering isotopic reactions.

Isochron age correlation lines were built using a best-fit regression approach. Combinations of accepted vs rejected points are automatically tried, in order to keep the MSWD down to its lowest value in which it does not change and at the same time maintaining the most sample steps possible and preferably consecutive steps. Any MSWD less than 2.0 is considered a valid isochron.

6.5.5 Results and Discussion of Results

The reported isotope correlation and age spectra results shows that the samples exhibit a trapped isotopic $^{40}Ar/^{36}Ar$ that is usually slightly greater than 295.5 (atmospheric), perhaps indicating that original trapped argon compositions may include unsupported ^{40}Ar (see Figs. 6.5, 6.6, and 6.7). Due to this, isochron ages were preferred over plateau ages because the former do not make any assumption about the trapped, initial argon composition but instead give a direct measurement of the trapped argon composition and thus probably making a better choice for the age values.

The $^{40}Ar/^{39}Ar$ geochronology yielded results spanning 11.9–0.30 Ma. Table 6.12 summarizes the results, while isotope correlation and age spectra can be found in Figs. 6.5, 6.6, and 6.7.

6.5.5.1 Santiago

In Santiago, and for the Flamengos Formation, samples ST25 and ST26 yielded respectively 3.99 ± 0.30 and 4.15 ± 0.20 Ma. These ages are in general agreement

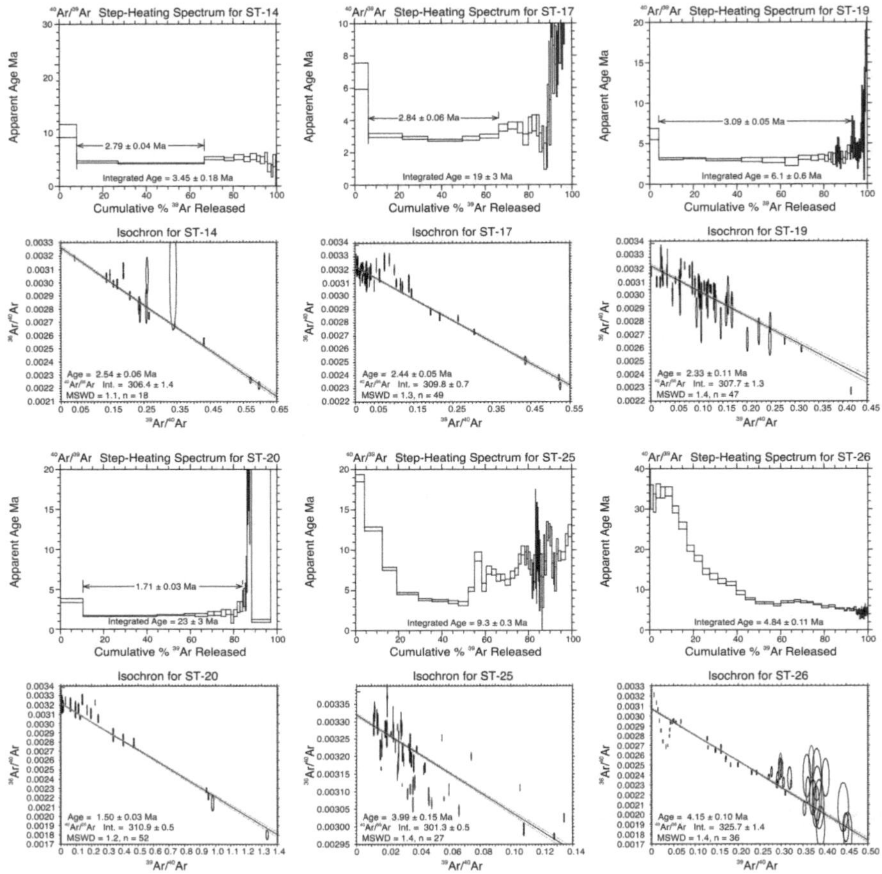

Fig. 6.5 Isotope correlation and age spectra (for comparison) for Santiago's lavas. The isochron ages with uncertainties (indicated) are calculated from the best fitting lines through collinear step compositions (*red ellipses* correspond to points discarded in the correlation). Heights of *rectangular boxes* on age spectra indicate estimated analytical error ($\pm 2\sigma$) for each step

with the ~4.6 Ma age proposed by Holm et al. [41] for the same unit. The samples collected in Pico da Antónia Eruptive Complex (ST14, ST17 and ST19), in the northern peninsula of Santiago, yielded ages between 2.33 and 2.54 Ma and lie in the 2.2–3.3 Ma interval proposed by Holm et al. [41] as the approximate extrusion interval for this sequence. However, sample ST20 collected in Porto Mosquito in a sequence attributed to the Pico da Antónia Eruptive Complex, yielded an age of 1.50 ± 0.08 Ma. This seems to extend the younger bound of Pico da Antónia Eruptive Complex to ~1.50 Ma, instead of the 2.2 Ma proposed by Holm et al. [41]. It is not impossible, however, for this sequence to belong to Assomada/Monte das Vacas Formation instead, but additional field work is needed to confirm this.

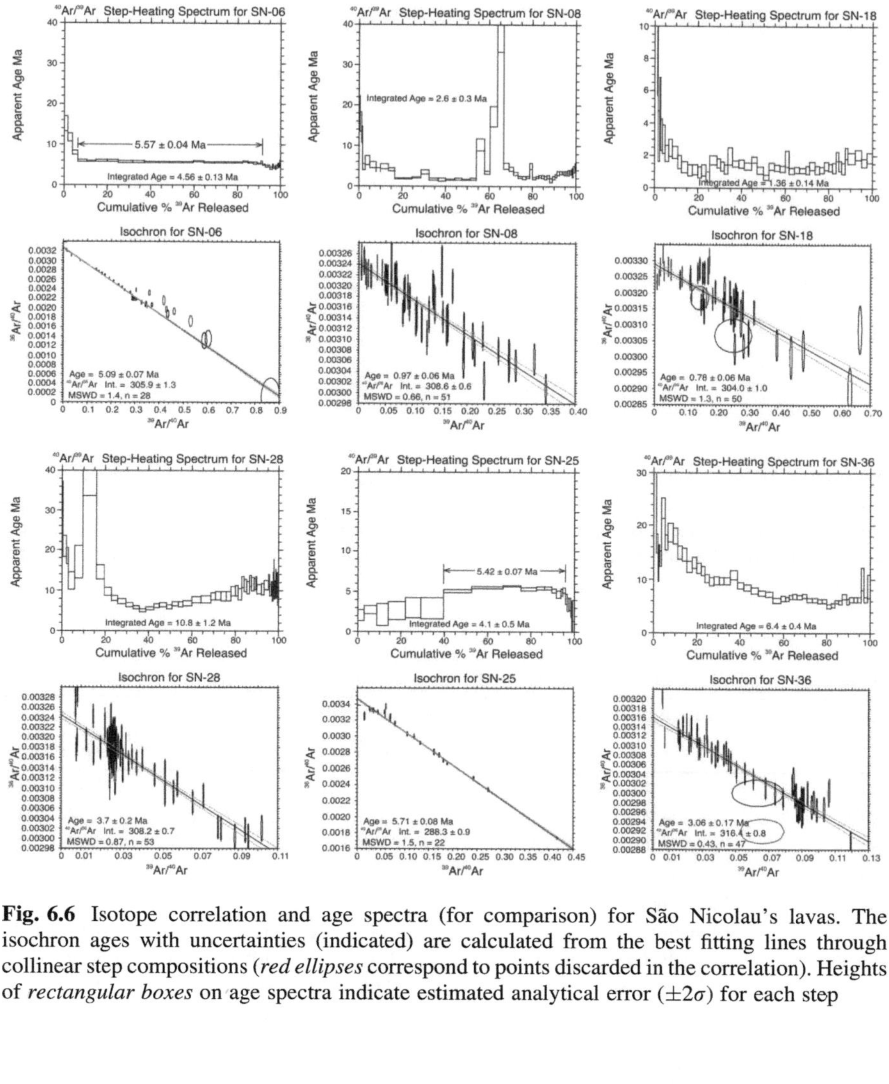

Fig. 6.6 Isotope correlation and age spectra (for comparison) for São Nicolau's lavas. The isochron ages with uncertainties (indicated) are calculated from the best fitting lines through collinear step compositions (*red ellipses* correspond to points discarded in the correlation). Heights of *rectangular boxes* on age spectra indicate estimated analytical error ($\pm 2\sigma$) for each step

6.5.5.2 São Nicolau

In São Nicolau our ages span from 0.78 to 5.71 Ma. Sample SN25, collected on the top of the Figueira de Coxe Formation, yielded an age of 5.71 ± 0.14 Ma and agrees very well with the >5.7 Ma younger bound proposed by Duprat et al. [25]. Samples SN06, SN23 and SN36 yielded ages between 5.09 and 3.06 Ma and are in rough agreement with the previously proposed ages for the Main Eruptive Complex, between ~4.7 and ~2.5 Ma (Stage 2 of Duprat et al. [25]. Finally, Samples SN08 and SN18 yielded 0.97 ± 0.12 and 0.78 ± 0.12 Ma, respectively. These are in general agreement with the ages obtained by Duprat et al. [25] in

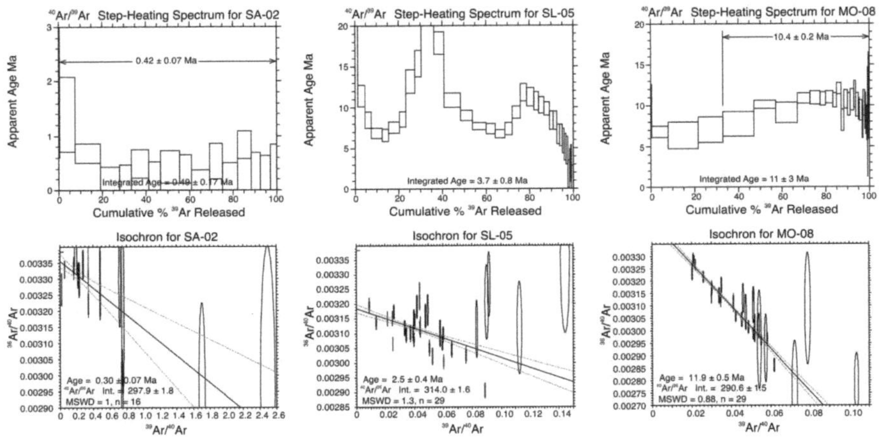

Fig. 6.7 Isotope correlation and age spectra (for comparison) for Santo Antão, Sal and Maio's lavas. The isochron ages with uncertainties (indicated) are calculated from the best fitting lines through collinear step compositions (*red ellipses* correspond to points discarded in the correlation). Heights of *rectangular boxes* on age spectra indicate estimated analytical error ($\pm 2\sigma$) for each step

Table 6.12 Summary of $^{40}Ar/^{39}Ar$ laser step-heating geochronology results

Sample ID	Stratigraphic unit	Isotope correlation			
		n	MSWD	$(^{40}Ar/^{39}Ar)_i$	Age
ST14	Pico da Antónia E. C.	18	1.1	306.4 ± 1.4	2.54 ± 0.06
ST17	Pico da Antónia E. C.	49	1.3	309.8 ± 0.7	2.44 ± 0.05
ST19	Pico da Antónia E. C.	47	1.4	307.7 ± 1.3	2.33 ± 0.11
ST20	Pico da Antónia E. C.	52	1.2	310.9 ± 0.5	1.50 ± 0.03
ST25	Flamengos Fm	27	1.4	301.3 ± 0.5	3.99 ± 0.15
ST26	Flamengos Fm	36	1.4	325.7 ± 1.4	4.15 ± 0.10
SN06	Main E. Complex	28	1.4	305.9 ± 1.3	5.09 ± 0.07
SN08	Preguiça Fm	51	0.66	308.6 ± 0.6	0.97 ± 0.06
SN18	Preguiça Fm	50	1.3	304.0 ± 1.0	0.78 ± 0.06
SN23	Main E. Complex	53	0.87	308.2 ± 0.7	3.7 ± 0.2
SN25	Fig. de Coxe Fm	22	1.5	288.3 ± 0.9	5.71 ± 0.08
SN36	Main E. Complex	47	0.43	316.4 ± 0.8	3.06 ± 0.17
SA02	Young Tarrafal V.U.	16	1.0	297.9 ± 1.8	0.30 ± 0.07
SL05	Serra Negra Fm	29	1.3	314.0 ± 1.6	2.5 ± 0.4
MO08	Casas Velhas Fm	29	0.88	290.6 ± 1.5	11.9 ± 0.5

n is the number of heating increments in the $^{40}Ar/^{39}Ar$ in the isotope correlation. MSWD is the mean square of weighted deviates resulting from the linear regression of the $^{39}Ar/^{40}Ar$ vs $^{36}Ar/^{40}Ar$ correlation. $(^{40}Ar/^{39}Ar)_i$ is the value of the non-radiogenic trapped argon determined from the isotope correlation. All errors are 1σ and do not include the error on the *J* value. The isotope correlation ages are considered the most reliable estimates for dating the emplacement of submarine volcanic units because they do not make any assumption about trapped, initial argon compositions and thus are preferred in this study nearby locations (between ~1.7 and ~1.0 Ma) and seem to support the inclusion of the younger lava deltas into the Preguiça Formation.

6.5.5.3 Santo Antão

Sample SA02 from Ponta do Atum yielded an age of 0.30 ± 0.14 Ma, in agreement with the ages published by Holm et al. [39] for the nearby valley-filling sequence of Young Tarrafal volcanic unit.

6.5.5.4 Sal

Sample SL05 yielded an age of 2.5 ± 0.8 Ma, a value significantly younger than the ~5.5 Ma reported by Torres et al. [42] and by Holm et al. [41] for the same formation.

6.5.5.5 Maio

Sample MO08 yielded an age of 11.9 ± 1.0 Ma. This age is in agreement with the conventional K/Ar ages published by Mitchell et al. [43] for the Casas Velhas Formation.

6.5.6 Summary

A laser step-heating $^{40}Ar/^{39}Ar$ geochronology method was successfully applied on whole rock samples. A total of 15 samples were dated, with a focus on the islands of Santiago and São Nicolau, and yielded ages spanning from ~11.9 to ~0.30 Ma.

6.6 Summary

One of the crucial steps in the reconstruction of the history of vertical movements affecting an ocean island lies in accurately determining the time at which the relative sea-level was at a particular position. Thus, the quality of such reconstructions depends greatly upon the quality of the geochronological methods available to date the past sea-level markers used in the reconstructions. For the present study we attempted to date our markers through three different isotopic systems and through biostratigraphy. From the three isotopic methods used in this study, $^{40}Ar/^{39}Ar$ geochronology proved to be the most reliable and versatile. This method enabled the construction of a reasonable age frame (mostly focused in the islands of Santiago and São Nicolau) from which vertical movements can be inferred. It also contributed to our knowledge of the volcanostratigraphy of some of the islands in the study. The effectiveness of U/Th in our studies is somewhat compromised by the small age range it is applicable to. Nonetheless, it allowed the dating of some the Quaternary terraces of São Nicolau and Brava that would remain otherwise undated. Finally and disappointingly strontium isotope stratigraphy proved to be unreliable in the Cape Verde sediments for several possible reasons. These are probably due to pervasive alteration under the semi-arid conditions of the archipelago, or due to possible hydrothermal alteration induced by the emplacement of submarine lavas atop the sediments. Biostratigraphy proved essential in the study of pelagic carbonates, but it lacks the necessary resolution when it comes to macrofossils.

References

1. Elderfield, H. (1986). Strontium isotope stratigraphy. *Palaeogeography, Palaeoclimatology, Palaeoecology, 57*(1), 71–90.
2. Veizer, J. (1989). Strontium isotopes in seawater through time. *Annual Review of Earth and Planetary Sciences, 17*(1), 141–167.
3. McArthur, J. (1994). Recent trends in strontium isotope stratigraphy. *Terra Nova, 6*(4), 331–358.
4. McArthur, J., Howarth, R., & Bailey, T. (2001). Strontium isotope stratigraphy: LOWESS Version 3: Best fit to the marine Sr-isotope curve for 0–509 Ma and accompanying look-up table for deriving numerical age. *The Journal of Geology, 109*(2), 155–170.
5. McArthur, J., Donovan, D., Thirlwall, M., Fouke, B., & Mattey, D. (2000). Strontium isotope profile of the early Toarcian (Jurassic) oceanic anoxic event, the duration of ammonite biozones, and belemnite palaeotemperatures. *Earth and Planetary Science Letters, 179*(2), 269–285.
6. Edwards, R., Gallup, C., & Cheng, H. (2003). Uranium-series dating of marine and lacustrine carbonates. *Reviews in Mineralogy and Geochemistry, 52*(1), 363–405.
7. McDougall, I., & Harrison, T. (1999). *Geochronology and thermochronology by the 40Ar/39Ar method*. New York: Oxford University Press.
8. Hess, J., Bender, M., & Schilling, J. (1986). Evolution of the ratio of strontium-87 to strontium-86 in seawater from Cretaceous to present. *Science, 231*(4741), 979–984.
9. Howarth, R., & McArthur, J. (1997). Statistics for strontium isotope stratigraphy: a robust LOWESS fit to the marine Sr-isotope curve for 0 to 206 Ma, with look-up table for derivation of numeric age. *The Journal of Geology, 105*(4), 441–456.
10. Veizer, J., Buhl, D., Diener, A., Ebneth, S., Podlaha, O., Bruckschen, P., et al. (1997). Strontium isotope stratigraphy: Potential resolution and event correlation. *Palaeogeography Palaeoclimatology Palaeoecology, 132*(1), 65–78.
11. Veizer, J., Ala, D., Azmy, K., Bruckschen, P., Buhl, D., Bruhn, F., et al. (1999). 87Sr/86Sr, δ 13C and δ 18O evolution of Phanerozoic seawater. *Chemical Geology, 161*(1–3), 59–88.
12. Vance, D., Teagle, D., & Foster, G. (2009). Variable Quaternary chemical weathering fluxes and imbalances in marine geochemical budgets. *Nature, 458*(7237), 493–496.
13. Dingle, R., McArthur, J., & Vroon, P. (1997). Oligocene and Pliocene interglacial events in the Antarctic Peninsula dated using strontium isotope stratigraphy. *Journal of the Geological Society, 154*(2), 257–264.
14. Bailey, T., McArthur, J., Prince, H., & Thirlwall, M. (2000). Dissolution methods for strontium isotope stratigraphy: whole rock analysis. *Chemical Geology, 167*(3–4), 313–319.

15. Lu, F. (2008). Pristine or altered: low-Mg calcite shells survived from massive dolomitization? A case study from Miocene carbonates. *Geo-Marine Letters, 28*(5), 339–349.
16. McArthur, J., Kennedy, W., Chen, M., Thirlwall, M., & Gale, A. (1994). Strontium isotope stratigraphy for Late Cretaceous time: direct numerical calibration of the Sr isotope curve based on the US Western Interior. *Palaeogeography, Palaeoclimatology, Palaeoecology, 108*(1–2), 95–119.
17. Veizer, J. (1983). Trace elements and isotopes in sedimentary carbonates. *Reviews in Mineralogy and Geochemistry, 11*(1), 265–299.
18. Morrison, J., & Brand, U. (1988). An evaluation of diagenesis and chemostratigraphy of Upper Cretaceous molluscs from the Canadian Interior Seaway. *Chemical Geology, 72*(3), 235–248.
19. Brand, U. (1991). Strontium isotope diagenesis of biogenic aragonite and low-Mg calcite. *Geochimica et Cosmochimica Acta, 55*(2), 505–513.
20. Stoll, H., & Schrag, D. (1998). Effects of Quaternary sea level cycles on strontium in seawater. *Geochimica et Cosmochimica Acta, 62*(7), 1107–1118.
21. Lorrain, A., Gillikin, D., Paulet, Y., Chauvaud, L., Le Mercier, A., Navez, J., & Andre, L. (2005). Strong kinetic effects on Sr/Ca ratios in the calcitic bivalve Pecten maximus. *Geology, 33*(12), 965–968.
22. Bernoulli, D., Hottinger, L., Spezzaferri, S., & Stille, P. (2007). Miocene shallow-water limestones from São Nicolau (Cabo Verde): Caribbean-type benthic fauna and time constraints for volcanism. *Swiss Journal of Geosciences, 100*(2), 215–225.
23. Serralheiro, A., & Ubaldo, M. (1979). Estudo estratigráfico dos sedimentos do Campo da Preguiça ilha de S. Nicolau (Cabo Verde). *Garcia de Orta, Serviços Geológicos, 3*(1–2), 75–82.
24. Macedo, J., Serralheiro, A., & Silva, L. (1988). Notícia Explicativa da Carta Geológica da Ilha de S Nicolau (Cabo Verde) na escala de 1:50000. *Garcia de Orta, Serviços Geológicos, 11*(1–2), 1–32.
25. Duprat, H., Friis, J., Holm, P., Grandvuinet, T., & Sørensen, R. (2007). The volcanic and geochemical development of São Nicolau, Cape Verde Islands: Constraints from field and 40Ar/39Ar evidence. *Journal of Volcanology and Geothermal Research, 162*(1–2), 1–19.
26. Sun, S., & McDonough, W. (1989). Chemical and isotopic systematics of oceanic basalts: implications for mantle composition and processes. *Geological Society London Special Publications, 42*(1), 313–345.
27. Davies, G., Norry, M., Gerlach, D., & Cliff, R. (1989). A combined chemical and Pb-Sr-Nd isotope study of the Azores and Cape Verde hot spots; the geodynamic implications. *Geological Society of London Special Publications, 42*(1), 231–255.
28. Bourdon, B., Turner, S., Henderson, G., & Lundstrom, C. (2003). Introduction to U-series geochemistry. *Reviews in Mineralogy and Geochemistry, 52*(1), 1–21.
29. Stirling, C., Lee, D., Christensen, J., & Halliday, A. (2000). High-precision in situ 238U–234U–230Th isotopic analysis using laser ablation multiple-collector ICPMS. *Geochimica et Cosmochimica Acta, 64*(21), 3737–3750.
30. Hoffmann, D., Spotl, C., & Mangini, A. (2009). Micromill and in situ laser ablation sampling techniques for high spatial resolution MC-ICPMS U-Th dating of carbonates. *Chemical Geology, 259*(3–4), 253–261.
31. Stirling, C., Esat, T., Lambeck, K., & McCulloch, M. (1998). Timing and duration of the Last Interglacial: evidence for a restricted interval of widespread coral reef growth. *Earth and Planetary Science Letters, 160*(3–4), 745–762.
32. Esat, T., & Yokoyama, Y. (2006). Variability in the uranium isotopic composition of the oceans over glacial–interglacial timescales. *Geochimica et Cosmochimica Acta, 70*(16), 4140–4150.
33. Renne, P., Swisher, C., Deino, A., Karner, D., Owens, T., & DePaolo, D. (1998). Intercalibration of standards, absolute ages and uncertainties in 40Ar/39Ar dating. *Chemical Geology, 145*(1–2), 117–152.

34. Kelley, S. (1995). Ar-Ar dating by laser microprobe. In P. Potts, F. Bowles, S. Reed, & M. Cave (Eds.), *Microprobe techniques in the earth sciences,* chap. 8 (pp. 327–358). London, UK: Chapman & Hall.
35. Kelley, S. (2002). Excess argon in K–Ar and Ar–Ar geochronology. *Chemical Geology, 188*(1–2), 1–22.
36. Serralheiro, A. (1976). A Geologia da Ilha de Santiago (Cabo Verde). *Boletim do Museu e Laboratorio Mineralógico e Geológico da Faculdade de Ciências, 14,* 157–369.
37. Silva, L., Le Bas, M., & Robertson, A. (1981). An oceanic carbonatite volcano on Santiago, Cape Verde Islands. *Nature, 294*(5842), 644–645.
38. Ramalho, R., Helffrich, G., Schmidt, D., & Vance, D. (2010). Tracers of uplift and subsidence in the Cape Verde Archipelago. *Journal of the Geological Society, 167*(3), 519–538.
39. Holm, P., Wilson, J. R., Christensen, B., Hansen, L., Hansen, S., Khein, K. M., et al. (2006). Sampling the Cape Verde plume: Evolution of melt compositions on Santo Antão, Cape Verde Islands. *Journal of Petrology, 47*(1), 145–189.
40. Kuiper, K., Deino, A., Hilgen, F., Krijgsman, W., Renne, P., Wijbrans, J. (2008). Synchronizing rock clocks of Earth history. *Science, 320*(5875), 500–504.
41. Holm, P., Grandvuinet, T., Friis, J., Wilson, J. R., Barker, A. K., & Plesner, S. (2008). An 40Ar-39Ar study of the Cape Verde hot spot: Temporal evolution in a semistationary plate environment. *Journal of Geophysical Research (Solid Earth), 113*(B8), B08201.
42. Torres, P., Silva, L., Serralheiro, A., Tassinari, C., & Munhá, J. (2002). Enquadramento geocronológico pelo método K/Ar das principais sequências vulcano-estratigráficas da Ilha do Sal - Cabo Verde. *Garcia de Orta. Serviços Geológicos, 18*(1–2), 9–13.
43. Mitchell, J., Bas, M. L., Zielonka, J., & Furnes, H. (1983). On dating the magmatism of Maio, Cape Verde Islands. *Earth and Planetary Science Letters, 64*(1), 61–76.
44. Bernard-Griffiths, J., Cantagrel, J.-M., Alves, C., Mendes, F., Serralheiro, A., & Macedo, J. (1975). Geochronologie: Donnés radiometriques potassium-argon sur quelques formations magmatiques des îles de l'archipel du Cap Vert CR. *Seances Academy of Science Series D, 280,* 2429–2432.
45. Plesner, S., Holm, P. M., & Wilson, J. R. (2002). 40Ar-39Ar geochronology of Santo Antão, Cape Verde Islands. *Journal of Volcanology and Geothermal Research, 120*(1–2), 103–121.

Chapter 7
Vertical Movements of Ocean Island Volcanoes: Insights from a Stationary Plate

7.1 Introduction

Ocean island volcanoes are subject to vertical movements during their lifetime. Darwin's [1] model for atoll formation implies a process leading to island subsidence [2]. However, the perception that the ocean bottom and the islands above it undergo vertical movements gained a physical footing when scientists realised that oceanic plates experience thermal subsidence [3–5] and that volcanic edifices cause plate bending by surface loading [6–10]. The discovery of large areas of the ocean floor that surround sites of mid-plate volcanism and that exhibit significantly shallower-than-expected depths, known as "swells" [11, 12], led to the realization that not only does subsidence occur but significant uplift could potentially ensue as well. Likewise, the presence of apparently uplifted atolls was soon linked to the existence of asthenospheric (and lithospheric) anomalies [2, 9] or related to far-field effects of island loading [10]. The general view, mostly based on the prototypical Hawaiian model, is that islands are formed atop or on the frontal slope of hotspot swells and subsequently subside under their own weight, creating a flexural moat. Towards the end of their main eruptive life, and as they move away from the melting source, islands gradually start to experience uplift as they pass over the flexural bulge created by new islands forming upstream. Erosion and mass wasting events that redistribute the load over a larger area also cause uplift. Older islands, as they move away from the flexural bulge, start to re-subside, forming atolls and eventually disappearing under the sea's surface, becoming guyots. This view is strictly valid for island chains on fast moving plates. What happens on stationary plates, when the islands' positions are fixed relative to the melting source? What vertical movements do they experience and what are the mechanisms that control these movements? To explore these issues, we studied the Cape Verde Archipelago, a group of islands that is approximately stationary in the hotspot reference frame [13–15] and that lies on top of the largest bathymetric anomaly in the oceans: the Cape Verde Rise [16].

R. A. S. Ramalho, *Building the Cape Verde Islands*, Springer Theses,
DOI: 10.1007/978-3-642-19103-9_7, © Springer-Verlag Berlin Heidelberg 2011

The vertical movements that affect ocean islands probably result from a combination of several processes, chiefly: (1) the ageing of the underlying lithosphere [3, 4]; (2) hotspot swell development and dynamics [9, 14, 15, 17, 18]; (3) lithospheric flexural response to the mass transfer associated with island volcanism that causes surface and subsurface loading [7–10, 14]; (4) and the subsequent redistribution of volume by erosion and mass-wasting effects [19–21].

Consider first the effects of lithospheric ageing. Oceanic lithosphere is produced at mid-ocean ridges and subsequently cools with age, causing thermal subsidence. The process is predictable: a simple equation describes the expected depth of the plate based on its age [4, 5]. The depth/age relationship follows a declining trend that tends to vertical stability when the lithosphere reaches 80–100 Ma, asymptotically at a depth of ~5,500 m [4, 5]. By using the extreme values of the curve, one can infer the maximum amount of subsidence the ocean basins experience through cooling: ~2,500 m. Consequently, ocean islands built on young lithosphere will appear to subside significantly if only plate cooling is considered.

The origin of hotspot swells is still controversial and their development is at present poorly constrained and strongly model-dependent. However, since islands are built on top of these structures, the role of swell development in island uplift/subsidence should be assessed. The fact that swells persist attests to their isostatic balance [22]. Notwithstanding their persistence, we assume that hotspot swells are created when the vertical buoyancy flux first impinges on the lithosphere, resulting in a very fast initial uplift (creating the swell's full height in less than 1–2 Ma) followed by slow subsidence or stability depending on the plate's velocity relative to the melting source [14, 15, 17, 18]. In fast moving plates the swell should decay since the plate moves away from the main source of buoyancy, while in stationary plates vertical stability is expected due to their fixity relative to the buoyancy source. Data from the Hawaiian chain suggests average subsidence rates of 20–30 m/Ma until stability is reached around 50 Ma after the island has passed the swell's crest [17].

Ocean islands are constituted of prominent edifices created by mass transfer to the surface—a process that causes significant vertical displacement of the islands themselves. The underlying lithosphere responds to the mass of the volcano by bending, creating a deformation field that extends well beyond the load itself [17] of both subsidence (the flexural depression or moat) and uplift (the forebulge or upwarp) [23]. The amplitude of deflection created by island loading typically ranges from 1 to 3 km, partially or totally compensating the swell height, and the radius of deflection normally ranges between 80 km (young/thin lithosphere) and 230 km (old/thick lithosphere) [7]. The forebulge extends outwards to a similar length from the moat and its maximum height corresponds to ~4% of the moat's depth [23, 24]. The lithospheric response time to loading is fast (<1 Ma) and is practically synchronous with the building of the bulk of the edifice, so volcanoes should be fully compensated almost as soon as they are built [7, 17]. However, since surface loading causes deformation over a much broader area than the load itself, islands in the vicinity are differentially affected depending on the distance from the load [10, 14]. Thus, nearby islands will undergo vertical displacements that vary from significant subsidence to slight uplift depending on their position, and cumulative effects from multiple loading may occur. Despite these complexities, the uplift component of cumulative far-field surface loading effects seems to be no greater than 100–120 m. In contrast, considerable uplift may be caused by subsurface loading [23]. This process is frequently associated with hotspot activity and occurs when the ponding of magmatic material at the base of the crust acts as an upward load, leading to uplift [23]. Subsurface loading is normally restricted to the edifice vicinity and may vary considerably from edifice to edifice.

Once the bulk of a volcanic edifice is built, the slow process of mass redistribution starts. Young islands are prominent and unstable features liable to erosion and flank collapse, resulting in the deposition of sedimentary aprons [21]. These sedimentary accumulations fill in the moat and may cover a large area around the edifice. Consequently, flexural adjustments to this mass redistribution are expected to occur, resulting in uplift.

By understanding the history of vertical movements affecting ocean islands in both fast-moving and stationary plate environments, one can assess the individual contribution of the various mechanisms and gain insight into the processes that cause them. A way to trace those movements is to look for markers of past sea-level in the islands' geological record, and to correlate their present position with the sea-level height contemporaneous with their formation. A remarkable feature of the Cape Verde Archipelago is the relative abundance of sea-level palaeo-markers in the islands' stratigraphical sequence, making it an ideal place to research island freeboard [25]. We have identified a number of strategic locations on the island of Santiago (in the southern chain), and on the island of São Nicolau (in the northern chain), to study and sample in detail. Here we present our vertical reconstructions that show significant differential uplift and test the possible contribution of each mechanism in the uplift process. Based on our calculations, we propose a local rather than regional mechanism to be responsible for significant island uplift.

7.2 Geological Background

7.2.1 The Cape Verde Archipelago

The Cape Verde Archipelago is a group of ten volcanic islands situated 450–600 km off the west coast of Africa, 15°–17° north of the equator (see Fig. 7.1). The archipelago is considered to be the result of a hotspot whose activity spans at least ~26 Ma [27]. Since the archipelago lies on the African plate, considered to be quasi-stationary in the hotspot reference frame [13], it is stationary relative to the melting source. The islands themselves lie on 120–140 Ma sea-floor [28] and seem to form two chains disposed in a rough semi-arc clustered slightly off-centre to the southwest of the bathymetric swell known as the Cape Verde Rise [14, 29].

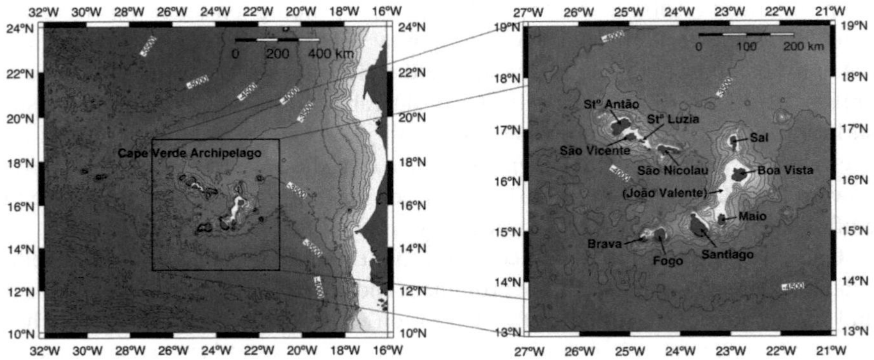

Fig. 7.1 Map of the Cape Verde Archipelago within the Cape Verde bathymetric swell, and detail. Note the existence of a "northern chain" comprising the islands of Santo Antão, São Vicente, Santa Luzia and islets, and São Nicolau, and a "southern chain" comprising Sal, Boa Vista, João Valente (a guyot), Maio, Santiago, Fogo and Brava. Bathymetric contours in meters

The swell is 1,400–1,600 km wide and $\sim$2.2 km high and the region is associated with a positive geoid anomaly of 8–12 m [14, 29]. The origins of the swell are still controversial but it seems it formed rapidly approximately 25 Ma ago since very few turbidites from the African continental margin have reached the top of rise since then [30, 31]. The crustal and mantle structure beneath the islands seems to be characterised by an unevenly thickened crust (up to 22 km) underlain by a 60–80 km thick region of faster-than-average seismic velocities interpreted as a depleted swell root, residual from partial melting [32]. The crustal thickening seems to be confined to the islands' region [33, 34], and the subsurface load associated with this thickening might be causing a regional uplift (of $\sim$400 m) that restricts the amplitude of the flexural moat to just 1,600 m (instead of the expected 2 km inflection) and tilts the moat infill outwards from the islands [33].

The exposed geology of the Cape Verdes is complex and is still imperfectly known. There is no evident hotspot track but there is possibly an age progression in the southern chain, from east (older and denuded islands) to west (younger and prominent islands), as suggested by the geomorphology of the edifices and the age of the oldest exposed lithologies (see Fig. 7.2). Volcanism seems to be mostly alkaline, frequently undersaturated, and intrusive and extrusive carbonatites are common [40]. Due to the islands' stationary position relative to the melting source, their volcanic lifetime is normally long, characterized by several periods of volcanism with decreasing volumetric output and intercalated with long (1–4 Ma) periods of quiescence (see Fig. 7.2). Historical volcanism (<500 a) is only known on Fogo, whose latest eruption occurred in 1995, but most islands exhibit products erupted during the Quaternary.

The presence of eroded basement complexes, with uplifted portions of ocean-floor [43–47] and large pockets of plutonic rocks [48], denotes early cycles of volcanism that normally precede the main shield-building stage and are still poorly known. These might correspond to relics of the seamount and/or emergent island

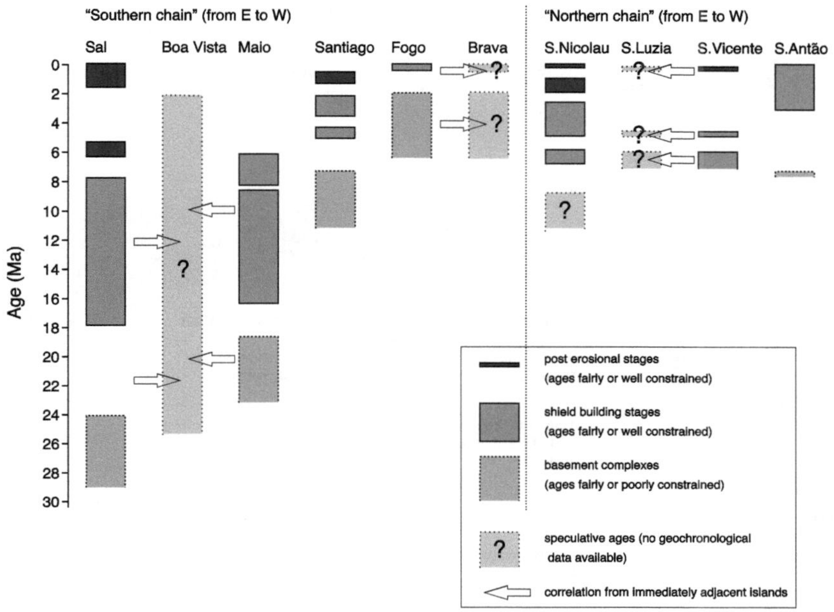

Fig. 7.2 Age distribution of exposed volcanic/intrusive products across the Cape Verde Archipelago. Modified from Holm et al. [35] and based on the geochonological data of Griffiths et al. [36], Mitchell et al. [37], Torres et al. [27], Plesner et al. [38], Jørgensen & Holm [39], Duprat et al. [40], Holm et al. [35], Foeken et al. [41]. Note the possible age progression of the oldest exposed rocks in the southern chain and the approximately synchronous extrusion of all the islands composing the northern chain

stages and may attest to significant vertical movements. However, their advanced state of alteration and the profusion of dyke swarms makes their study extremely difficult and discourages dating. Outstanding examples of these basal lithologies include Maio's old sequence, with Jurassic MORB basalts and Jurassic/Cretaceous deep-sea sediments, folded, faulted and interpreted as uplifted sea-floor possibly associated with the intrusion of the island's Central Igneous Complex [43, 45], and Sal's Ancient Eruptive Complex, a submarine volcanic sequence with intercalated pelagic sediments [25, 49–51].

In Cape Verde, in contrast with many other archipelagos, evidence for significant positive movement is not restricted to the basement complexes. The relative abundance of sea-level palaeo-markers extends throughout the stratigraphic sequence of many islands, suggesting significant differential uplift throughout the archipelago [25]. The islands of Santiago and São Nicolau are two noteworthy examples of seemingly uplifted edifices, while, for example, Santo Antão and São Vicente seem not to have experienced uplift [25]. For this paper we have focused our study on Santiago and São Nicolau, not only because of their unique geological records but also due to their central position in the archipelago and evolutionary similarities (see Figs. 7.2 and 7.3). Figure 7.3 compares the volcanostratigraphy of

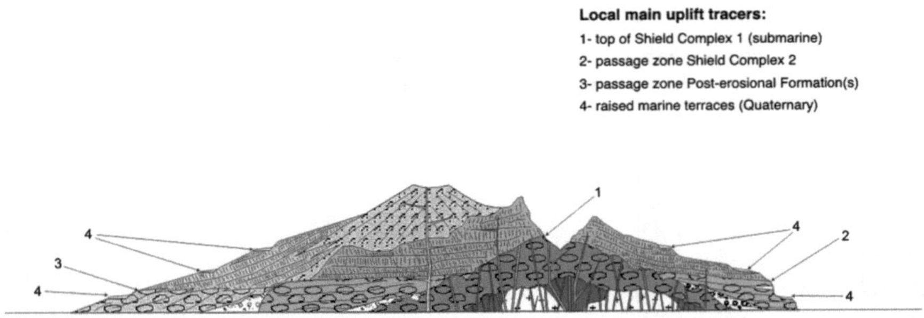

	Santiago	palaeo sea-level height	São Nicolau	palaeo sea-level height
marine terraces (Quaternary)	**marine terraces (Quaternary)**	≤ 100 m asl	**marine terraces (Quaternary)**	≤ 50 m asl
Post-erosional Formation(s)				
subaerial volcanics (piroclasts)	**Assomada/Monte das Vacas Formation(s) (only subaerial)**	≤ 40 m asl	**Monte Gordo Formation**	≤ 5 masl
subaerial volcanics (flows)			**Preguiça Formation**	≤ 40 m asl
submarine volcanics				
sediments				
Shield Complex 2 (subaerial shield-building stage)				
subaerial volcanics	**Pico da Antónia Volcanic Complex**	≤ 270 m asl	**Main Eruptive Complex**	≤ 100 m asl
submarine volcanics				
sediments				
Shield Complex 1 (submarine shield-building stage)				
submarine volcanics	**Flamengos Formation**	≥ 450 m asl	**Figueira de Coxe Formation**	≥ 270 m asl
sediments				
Basement Complex (relics of seamount stage?)				
intrusive	**Ancient Eruptive Complex**	(unknown)	**Ancient Eruptive Complex**	(unknown)
extrusive				

Fig. 7.3 Schematic representation of the main local uplift tracers (sea-level palaeo markers) in Santiago and São Nicolau and their relative stratigraphic position. Formation names and palaeo sea-level information for each tracer is shown. Santiago and São Nicolau's stratigraphic schemes after Serralheiro [48] and Macedo et al. [52], respectively

these two islands, with information regarding the main uplift tracers found on the island, including their present elevation.

7.2.2 *Santiago*

Santiago is an elongated shield volcano in an early post-erosional stage whose eruptive life spanned the Miocene(?) to the Quaternary [35]. The island is the largest in the archipelago and its shield morphology is still perceptible despite being cut by deep valleys and truncated by marine erosion. The island volcanostratigraphy was described by Serralheiro [48] and comprises, from the base to the top:

- *Ancient Eruptive Complex* This unit corresponds to the oldest exposed sequence and is profoundly weathered and eroded. It mostly comprises volcanic products of possible submarine origin, pervasively intruded by several generations of dyke swarms and small bodies of plutonic rocks [48]. The existence of uplifted portions of MORB-type basalts within this complex was later confirmed [46, 53] and denotes processes of upheaval that are still poorly documented. Possible subaerial structures, mostly of phonolitic and carbonatitic composition, are also found as late products of this unit [48]. The age of the Ancient Eruptive Complex is poorly constrained, but conventional K-Ar analyses performed on biotites from carbonatites yield ages of 9.6 ± 0.6 and 8.5 ± 1.4 Ma [36].
- *Flamengos Formation* This unit corresponds to an early shield-building stage, exclusively submarine. Its sequence comprises pillow-lavas, hyaloclastites and pillow-breccias of basaltic composition extruded rapidly $\sim$4.6 Ma [35]. The products of the Flamengos Formation are deeply eroded but crop out from the present sea-level up to elevations of $\sim$450 m above sea-level (asl) [48].
- *Órgãos Formation* This unit is composed of thick conglomerate deposits of terrestrial and marine origin and is the result of a significant erosion period affecting the previous units [48].
- *Pico da Antónia Eruptive Complex* This unit defines the present general morphology of the island and corresponds to a second shield-building stage where subaerial products dominate [25]. It is composed of a lower submarine member and an upper subaerial member, with minor submarine morphologies on its fringes [25, 54]. The submarine products of the lower Pico da Antónia Eruptive Complex can be found up to $\sim$270 m asl. In contrast, the submarine products of the upper member normally crop out below 100 m asl. The Pico da Antónia Eruptive Complex was possibly extruded between 3.3 and 2.2 Ma [35].
- *Assomada/Monte das Vacas Formation(s)* Originally considered as two separated units by Serralheiro [48], the Assomada and Monte das Vacas Formations were later considered by Holm et al. [35] as part of the same event due to their similar age (1.3–0.7 Ma). The Assomada/Monte das Vacas Formation(s) corresponds to a post-erosional stage that partially infilled the eroded shield topography. However, its products apparently did not reach the sea.
- *Quaternary sediments* This unit comprises the modern sediments associated with the recent destruction of the edifice. Marine terraces and poorly developed wavecut surfaces can be found up to elevations of 80–100 m asl [48]. Their formation was partially synchronous with the Assomada/Monte das Vacas eruptions.

7.2.3 São Nicolau

São Nicolau is the easternmost island of the northern chain. The island seems to have been formed by recurrent fissure volcanism along a main WNW–ESE rift arm and a minor N–S rift arm. Volcanism spanned the Miocene(?) to the Quaternary

[52] and the edifice is presently in a post-erosional stage. Its volcanostratigraphy was described by Serralheiro [48] and later detailed by Macedo et al. [52], comprising from the base to the top:

- *Ancient Eruptive Complex* The oldest exposed sequence on the island, the Ancient Eruptive Complex is mostly composed of weathered submarine and possibly subaerial products, intensely intruded by a dyke swarm. The unit is deeply eroded, attested to by the presence of the irregular unconformity that separates this unit from the overlying sequences. Like many other islands of the archipelago, the last erupted materials included in this unit are of phonolitic nature [52]. The precise age of the Ancient Eruptive Complex is unknown since there are no geochronological data available. However, its minimum age is constrained by the overlying sediments.
- *Late Miocene Marine Sediments* This unit comprises shallow marine sediments aged between 11.8 and 5.8 Ma, or even possibly between 6.2 and 5.8 Ma [55].
- *Figueira de Coxe Formation* This unit corresponds to an early shield-building stage, exclusively submarine. It comprises pillow-lavas, hyaloclastites, pillow-breccias and submarine pyroclasts, apparently extruded between 6.2 and 5.7 Ma bp (Stage 1 of Duprat et al. [40]). The products of this unit conformably cover the marine sediments [52]. The top is eroded but the Figueira de Coxe lavas can be found up to 250–260 m asl.
- *Conglomerate-Breccia Deposits* This unit comprises marine and terrestrial sediments and corresponds to a period of quiescence and erosion, responsible for the destruction of the Figueira de Coxe volcanic pile.
- *Main Eruptive Complex* Interpreted as a second shield-building stage, this unit mostly comprises effusive subaerial products that reached the sea, forming a fringe of submarine lavas at the base of the edifice. The passage zone between the subaerial and the submarine products is frequently found up to 100 m asl. The unit was apparently extruded between 4.7 and 2.6 Ma ago (Stage 2 of Duprat et al. [40]).
- *Preguiça Formation* This unit corresponds to the first post-erosional stage and created prominent lava deltas on the coast of the island. Among these we highlight the Terra Chã lava-delta, in eastern São Nicolau, due to its extension and preserved structure. The unit was probably extruded between 1.7 and 1 Ma ago (Stage 3 of Duprat et al. [40]).
- *Monte Gordo Formation* is the second post-erosional volcanic unit, extruded <1 Ma ago (Stage 4 of Duprat et al. [40]). Most of the preserved strombolian cones found on the island and associated flows (like the Fajã flows) are found in this unit, along with rare submarine morphologies.
- *Quaternary Sediments* This unit comprises the modern sediments associated with the recent destruction of the edifice. The coastal geomorphology of São Nicolau is characterised by a discontinuous gently dipping surface at 0–50 m asl and interpreted as a polygenetic marine abrasion surface. This surface was often covered by the products of the Preguiça and Monte Gordo Formations, as well as by alluvial fan deposits. Shallow marine sediments are sometimes found

intercalated with the volcanic products or on top of the sequence. Despite flow-mantling by the volcanic products, wavecut surfaces and marine terraces are distinguished at 5–7, 10, 20–30 and 50 m asl [25, 52].

7.3 Data

7.3.1 Uplift Information Sources

In order to trace island uplift we identified several targets for dating. In this paper we present the dating results from a number of submarine lavas and shallow marine carbonate deposits. Submarine lavas were collected as close as possible to the transition to subaerial morphology, or as close as possible to the top of the pile when the unit was entirely submarine or when the passage zone was eroded. Special attention was given to lava deltas, since these constitute incontrovertible sea-level markers. The shallow marine deposits were sampled for fossil corals. The locations and geological context of the samples are presented in Table 7.1. Figures 7.4 and 7.5 show sample localities. Sampling details in Chap. 6.

7.4 Methods

7.4.1 Uplift Calculations

The vertical displacement expressed by each marker was traced using a simple model:

$$D_v = h + d - H \tag{7.1}$$

where D_v is vertical displacement, h is the present elevation, d the inferred palaeo-depth, and H is the contemporaneous sea-level height extracted from an appropriate eustatic curve. H is the mean sea-level height of the absolute age uncertainty interval [age $- 2\sigma$, age $+ 2\sigma$]. Palaeo-depth information was estimated using the fossil content or the vertical distance to the geological subaerial/submarine interface (see supplement for details). Present elevations were estimated using a barometric altimeter in conjunction with a 1/25,000 topographical map. The barometric altimeter was regularly calibrated to avoid diurnal drift (using points of known elevation like trigonometric stations or the present sea-level). Errors in the determination of the elevations were assessed to be around ± 1 m near known markers and ± 5 m for distant ones. We chose the curve from Miller et al. [56] as the eustatic reference because it is the only recent curve that extends throughout the Miocene. Errors in the eustatic are neglected because, where divergent, they differ by less than the sea-level change amplitude. Uplift rates were calculated

Table 7.1 Summary of sample information for Santiago (samples with ST suffix) and São Nicolau (samples with SN suffix)

Sample ID	Coordinates (UTM WGS84)	General location	Type	Stratigraphic unit	Elevation (m)	2σ (m)	Palaeo-depth (m)
ST14	26 p 209737 1696097	Ponta Furna	s.f.	PAEC	153	5	27
ST17	26 p 209158 1696222	Ponta Furna	s.f.	PAEC	151	5	29
ST19	26 p 205821 1697012	Ponta Moreia	s.f.	PAEC	131	5	49
ST20	26 p 210347 1655934	Porto Mosquito	s.f.	PAEC	76	2	3
ST25	26 p 212120 1675155	Portal da Furna	s.f.	FFm	422	5	23
ST26	26 p 212247 1675348	Portal da Furna	s.f.	FFm	416	5	29
SN06	26 Q 815935 1835773	Castilhiano	s.f.	MEC	84	5	18
SN08	26 Q 816213 1835814	Castilhiano	s.f.	PFm	39	2	1
SN15	26 Q 809931 1832439	Ponta Trancoso	t	Qsed	29	2	5
SN16	26 Q 809694 1832452	Ponta Trancoso	t	Qsed	25	2	5
SN18	26 Q 810325 1832793	Rib. Gombeza	sa.f.	PFm	15	1	0
SN23	26 Q 816340 1832690	Talhada	s.f.	MEC	92	3	10
SN25	26 Q 797674 1840562	Fig. Coxe	s.f.	FCFm	185	5	80
SN28	26 Q 794919 1841616	Coelha	t	Qsed	21	2	5
SN36	27 Q 789776 1831457	Rib. Pta Pataca	s.f.	MEC	23	2	70

s.f. submarine flows, *t* terrace, *sa.f.* subaerial flow, *PAEC* Pico da Antónia Eruptive Complex, *FFm* Flamengos Fm, *MEC* Main Eruptive Complex, *PFm* Preguiça Fm, *Qsed* Quaternary sediments, *FCFm* Figueira de Coxe Fm

using a weighted linear model using joint uncertainties of the components. Maximum uplift rates were calculated using data with inferred palaeo-depth estimates while minimum rates used a palaeo-depth of 0 m.

7.4.2 Laser Step Heating $^{40}Ar/^{39}Ar$ *Geochronology*

The $^{40}Ar/^{39}Ar$ analyses were performed at the Institute of Mineralogy and Geochemistry at the University of Lausanne. Approximately $1 mm^3$ of washed but

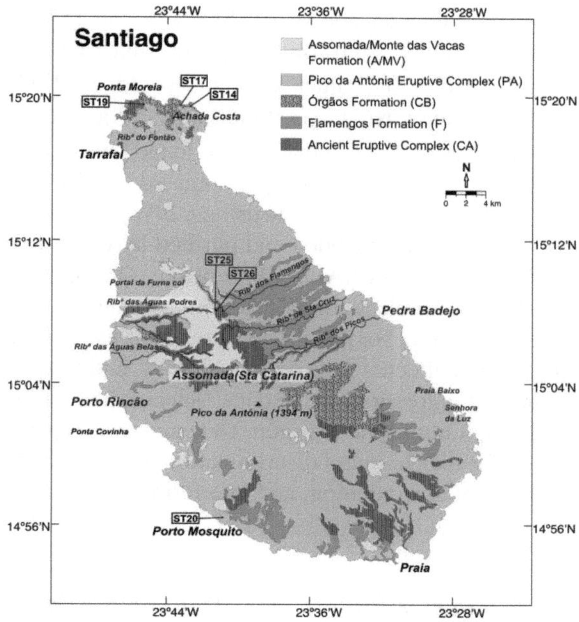

Fig. 7.4 Simplified geological map of Santiago with sample locations plotted. Map after Serralheiro [48]

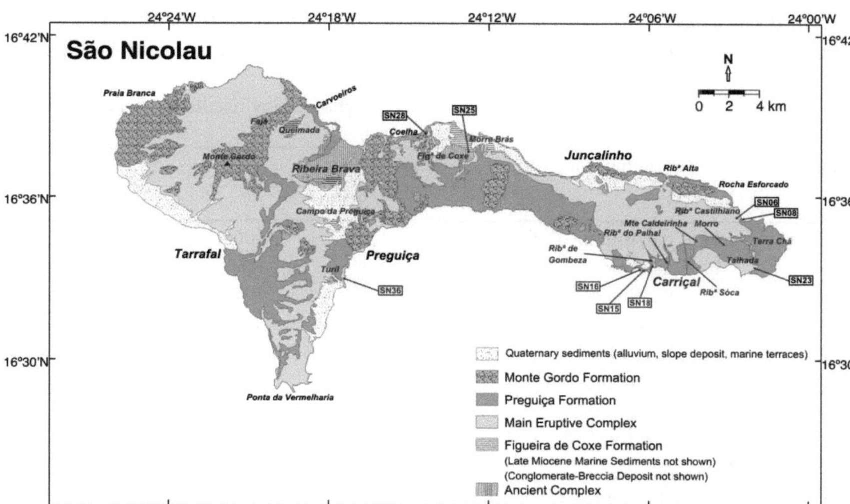

Fig. 7.5 Simplified geological map of São Nicolau with sample locations plotted. Map modified from Macedo et al. [52]

otherwise untreated whole rock sample, together with standards, were irradiated for 10 MWH with cadmium lining in the CLICIT facility at the Oregon State University TRIGA reactor. Monitoring of the neutron flux was done using Fish Canyon Tuff sanidine, using an age of 28.20 ± 0.08 Ma [57] and isotopic production ratios were determined from irradiated CaF_2 and KCl salts. For this irradiation, the following production values were measured: $^{(36/37)}Ca = 0.0002964 \pm 0.00000587$; $^{(39/37)}Ca = 0.00080 \pm 0.000053$; and $^{(38/39)}K = 0.0122 \pm 0.000028$. The irradiated cubes of basalt and standards were loaded into 3 mm wells within a stainless steel planchette attached to a fully automated extraction line evacuated to UHV conditions. Samples were degassed using a 20W CO_2 laser with a slightly defocused beam, and the gas was expanded and purified by exposure to a cold finger maintained at −132°C, and a hot SAES GP50 getter. The purified gas was expanded into a Nu Instruments Noblesse mass spectrometer and argon isotopes were measured using a Faraday detector for m/e 40, and ion counting multipliers for m/e 39, 38, 37, and 36. Data were collected for a period of 300 s, and time zero intercepts were determined by best-fit regressions to the data. Detector intercalibration was done using repeated analysis of atmospheric argon. Data and ages reported in Table 7.2 have been corrected for blanks, mass discrimination, detector intercalibration, radioactive decay subsequent to irradiation, and interfering isotopic reactions.

7.4.3 Laser Ablation MC-ICPMS U-Th Disequilibrium Analysis

Quaternary terraces were dated using laser ablation (LA) multi-collector (MC)—inductively coupled mass spectrometry (ICPMS) U-Th disequilibrium techniques

Table 7.2 Summary of $^{40}Ar/^{39}Ar$ laser step heating geochronology results for Santiago and São Nicolau

Sample ID	Stratigraphic unit	Isotope correlation			
		n	MSWD	$(^{40}Ar/^{39}Ar)_i$	Age
ST14	Pico da Antónia E. C.	18	1.1	306.4 ± 1.4	2.54 ± 0.06
ST17	Pico da Antónia E. C.	49	1.3	309.8 ± 0.7	2.44 ± 0.05
ST19	Pico da Antónia E. C.	47	1.4	307.7 ± 1.3	2.33 ± 0.11
ST20	Pico da Antónia E. C.	52	1.2	310.9 ± 0.5	1.50 ± 0.03
ST25	Flamengos Fm	27	1.4	301.3 ± 0.5	3.99 ± 0.15
ST26	Flamengos Fm	36	1.4	325.7 ± 1.4	4.15 ± 0.10
SN06	Main E. Complex	28	1.4	305.9 ± 1.3	5.09 ± 0.07
SN08	Preguiça Fm	51	0.66	308.6 ± 0.6	0.97 ± 0.06
SN18	Preguiça Fm	50	1.3	304.0 ± 1.0	0.78 ± 0.06
SN23	Main E. Complex	53	0.87	308.2 ± 0.7	3.7 ± 0.2
SN25	Fig. de Coxe Fm	22	1.5	288.3 ± 0.9	5.71 ± 0.08
SN36	Main E. Complex	47	0.43	316.4 ± 0.8	3.06 ± 0.17

n is the number of heating increments in the $^{40}Ar/^{39}Ar$ in the isotope correlation. MSWD is the mean square of weighted deviates resulting from the linear regression of the $^{39}Ar/^{40}Ar$ vs $^{36}Ar/^{40}Ar$ correlation. $(^{40}Ar/^{39}Ar)_i$ is the value of the non-radiogenic trapped argon determined from the isotope correlation. All errors are 1σ and do not include the error on the J value. The isotope correlation ages are considered the most reliable estimates for dating the emplacement of submarine volcanic units and thus are preferred in this study on fossil corals following the method outlined by Hoffmann et al. [58]. All MC-ICPMS measurements were performed in the Bristol Isotope Group (BIG) laboratory using a ThermoFinnigan Neptune coupled with a New Wave Research UP193HE ArF Excimer laser system. Coral samples were cut, polished, cleaned in an ultrasonic bath for 5 min and dried. Samples were then placed in a laser sample cell together with an internal carbonate U-Th LA calibration sample, which is a secular equilibrium calcite 'standard' for correction of instrumental biases of LA U-Th isotope measurements on $CaCO_3$. Potential matrix effects due to differences between aragonite and calcite are negligible within uncertainties achieved for U-Th isotope measurements using the LA technique [58]. Ablation is done using He as carrier gas which is mixed with Ar sample gas and N_2 in a quartz mixing cell before injection into the Ar plasma. Typical laser power density is 5 J/cm^2 at 70% power output, for U-Th isotope LA measurements presented in this study we used 7 Hz repetition rate and a 250 μm spot size. A LA measurement was done on a 0.5 mm long track, ablated by moving the laser spot at a speed of 20 μm/s along the track in six passes. A standard–sample–standard bracketing procedure was applied and data collection and corrections were done according to Hoffmann et al. [58].

7.5 Results

7.5.1 ^{40}Ar/^{39}Ar *Geochronology*

The ^{40}Ar/^{39}Ar ages span 0.78–5.71 Ma, providing a reasonable record for the relative position of sea-level through the recent period of island building. Table 7.2 summarizes the results.

7.5.1.1 Santiago

In Santiago, and for the Flamengos Formation, samples ST25 and ST26 yielded respectively 3.99 ± 0.30 and 4.15 ± 0.20 Ma. These ages are in general agreement with the ~4.6 Ma age proposed by Holm et al. [35] for the same unit. The samples collected in Pico da Antónia Eruptive Complex (ST14, ST17 and ST19), in the northern peninsula of Santiago, yielded ages between 2.33 and 2.54 Ma and lie in the 2.2–3.3 Ma interval proposed by Holm et al. [35] as the approximate extrusion interval for this sequence. However, sample ST20 collected in Porto Mosquito in a sequence attributed to the Pico da Antónia Eruptive Complex, yielded an age of 1.50 ± 0.08 Ma. This seems to extend the younger bound of Pico da Antónia Eruptive Complex to ~1.50 Ma, instead of the 2.2 Ma proposed by Holm et al. [35]. It is not impossible, however, for this sequence to belong to Assomada/Monte das Vacas Formation instead, but additional field work is needed to confirm this.

7.5.1.2 São Nicolau

In São Nicolau our ages span from 0.78 to 5.71 Ma. Sample SN25, collected on the top of the Figueira de Coxe Formation, yielded an age of 5.71 ± 0.14 Ma and agrees very well with the >5.7 Ma younger bound proposed by Duprat et al. [40]. Samples SN06, SN23 and SN36 yielded ages between 5.09 and 3.06 Ma and are in rough agreement with the previously proposed ages for the Main Eruptive Complex, between $\sim$4.70 and $\sim$2.5 Ma (Stage 2 of Duprat et al. [40]). Finally, Samples SN08 and SN18 yielded 0.97 ± 0.12 Ma and 0.78 ± 0.12 Ma, respectively. These are in general agreement with the ages obtained by Duprat et al. [40] in nearby locations (between $\sim$1.7 and $\sim$1 Ma) and seem to support the inclusion of the younger lava deltas into the Preguiça Formation.

7.5.2 *U-Th Geochronology*

The inferred ages for the three Quaternary terraces we sampled in São Nicolau are summarised in Table 7.3. The initial $^{234}U/^{238}U$ activity ratios agree within errors with the sea water value of 1.149 ± 0.002 [59, 60]. Perhaps surprisingly, samples SN16 and SN28 seem to have been formed during MIS 8, a low stand. Sample SN15 unfortunately presents an uncertainty too high to infer the MIS stage of formation. These three terraces exhibit older ages with increasing elevation.

7.5.3 *Uplift*

Figure 7.6 summarises the history of vertical movements affecting Santiago and São Nicolau islands. These reconstructions show that both Santiago and São Nicolau experienced significant uplift throughout the last 5–6 Ma. Each island exhibits a distinctive uplift trend: Santiago seems to exhibit a constant nearly-linear uplift of $\sim$100 m/Ma (0.1 mm/a) in the last 4 Ma, while São Nicolau seems to have been affected by considerable uplift at 5.7–5.1 Ma, and at 0.5 Ma–present, with a more modest average uplift rate of $\sim$32 m/Ma (0.032 mm/a). The total vertical displacement is in excess of $\sim$450 m for Santiago and in excess of $\sim$260 m for São Nicolau.

Table 7.3 U-Th disequilibrium ages for São Nicolau Quaternary terraces

Sample ID	$(^{230}Th/^{238}U)_A$	$(^{234}U/^{238}U)_A$	U-Th age (ka)	$(^{234}U/^{238}U)_{A0}$
SN15	1.049 ± 0.015	1.046 ± 0.005	$465.7 + 195 - 76$	$1.174 + 0.124 - 0.031$
SN16	1.012 ± 0.015	1.072 ± 0.004	283.6 ± 19.7	1.160 ± 0.010
SN28	0.988 ± 0.013	1.070 ± 0.006	258.3 ± 15.1	1.146 ± 0.011

7.6 Discussion

The uplift rates we find are comparable with other uplift rates found, for example, in the Hawaiian Islands: 0.032–0.1 mm/a in the Cape Verdes and 0.02–0.33 mm/a in Hawaii [61]. However, the latter study only encompassed the last 0.5 Ma whereas our study spans the last 5.7 Ma. Our rates are also possibly underestimated since the older palaeo-markers are considerably eroded.

It is perhaps surprising to note that the uplift trend experienced by Santiago and São Nicolau, with large resultant vertical displacements, is seemingly synchronous with the volcanic activity that built the bulk of their exposed edifices (correlate Fig. 7.6 with Figs. 7.2 and 7.3). Both islands seem to have been in fairly continuous uplift throughout their submarine and subaerial shield-building stages, extending well into their post-erosional stages. It is even more surprising to realise that nearby islands (especially in the northern chain) do not exhibit the same uplift trend, but are more stable [25]. This confirms that uplift is differential among the islands and not a regional feature.

What causes different uplift histories? Consider the individual contribution of each possible contributor to island uplift/subsidence.

First, the islands rest on sea-floor aged 120–140 Ma, an age beyond the asymptotic limit in bathymetric models [5]. Even if we consider a possible lithospheric thermal rejuvenation to an age of 61–81 Ma [33], the resulting uplift and subsequent subsidence is practically negligible. In addition, this mechanism should affect the islands uniformly.

Second, despite our poor understanding of the development of the Cape Verde Rise, its uplift seems to have happened in the early stages of hotspot activity, before the majority of the edifices were formed. Since the Cape Verdes are stationary relative to its melting source, we postulate that the buoyancy source that supports the swell has not waned with time, resulting in its vertical stability. Thus we assume that this mechanism had little effect on island uplift/subsidence and, if there was any effect, it also affected the islands in a uniform manner.

Surface loading is the next effect to consider. Assuming that: (1) both Santiago and São Nicolau did not subside significantly after the formation of their basement complexes, and (2) both edifices exert a downwards flexure by means of surface loading [33], we may infer that the subsidence associated with the flexural response to surface loading must have occurred in the early stages of island building when the bulk of the edifice was built and before the main shield-building stages. This is not surprising because the flexural response of the lithosphere to surface loading is relatively fast and the edifices reach stability soon after their bulk volume is extruded [7, 14, 17]. Thus, only the basement complexes are expected to record considerable direct loading subsidence.

Could the uplift trend be caused by cumulative far-field effects of multiple surface loading? Since the islands form a cluster whose inter-island distances are small (ranging 10–290 km), most islands lie in the flexural depression of nearby islands, or, if they were built synchronously, they acted as a combined load.

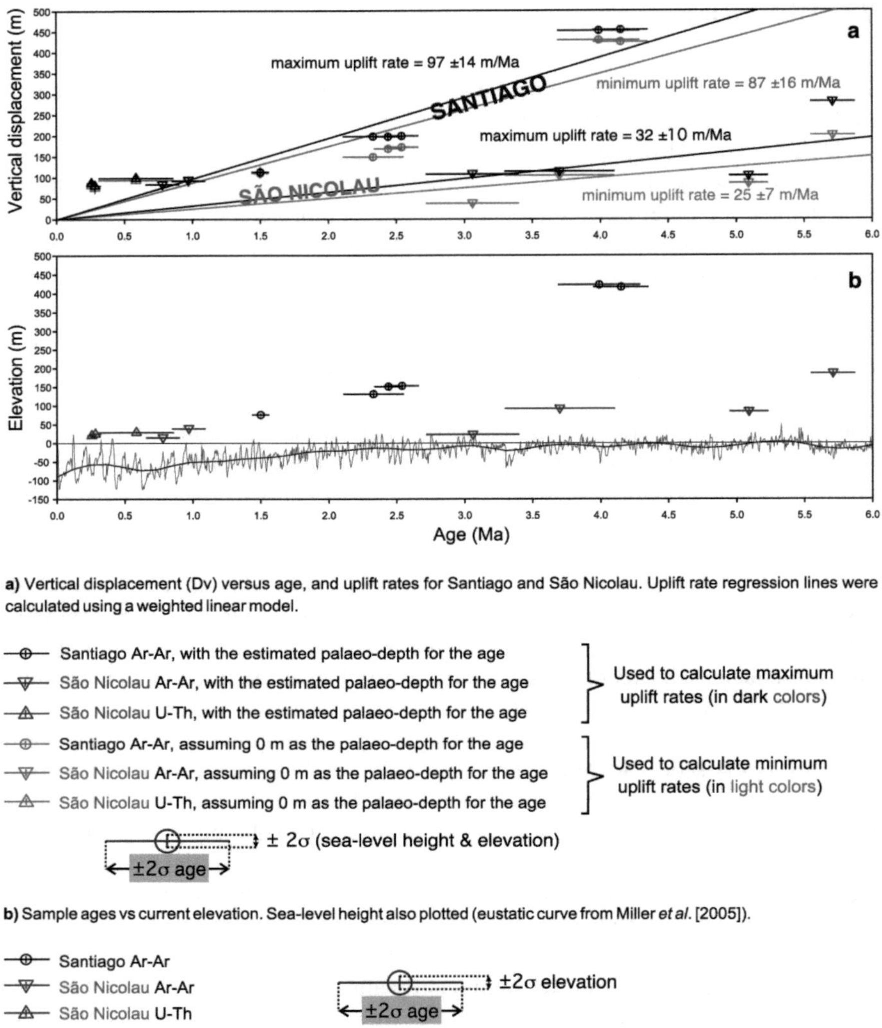

Fig. 7.6 Uplift reconstructions for Santiago and São Nicolau islands: **a** Vertical displacement and calculated uplift rates; **b** sample position regarding its age, its present elevation and the sea level height for the age (eustatic curve from Miller et al. [55]. Note that both islands exhibit an uplift trend, although at different rates and different timings

Nevertheless it is useful to try to reconstruct the history of progressive loading in the Cape Verdes. The age of onset of volcanism in most islands is poorly constrained, but using the published geochronological data (see Fig. 7.2) as well as to the geometry of the edifices, we postulate that archipelago was loaded during three main epochs: (1) extrusion of the Sal, Boa Vista, João Valente (now a guyot) and

Maio edifices 26–20 Ma ago; (2) extrusion of all the northern edifice (comprising Santo Antão, São Vicente, Santa Luzia and São Nicolau) and Santiago edifices 8–4 Ma ago; (3) extrusion of Fogo and Brava <4 Ma ago. Hence, the extrusion of the northern and Santiago edifices are likely to have caused subsidence in the Sal, Boa Vista, João Valente and Maio edifices 8–4 Ma ago. Likewise, the extrusion of Fogo and Brava are likely to have caused subsidence in all other islands except Santo Antão, Boa Vista and Sal, where slight uplift might have occurred and in São Vicente where neither subsidence or uplift is expected. Using the flexural model defined by Brotchie and Silvester [7] and for $T_e \sim 30$ km [32, 33], $E = 100$ GPa and $\rho_{\mathrm{mantle}} = 3,300$ kg/m^3, the uplift exerted on Sal by the loading of Fogo and Brava edifices is <20 m. Fogo and Brava should have experienced initial subsidence (when the edifices were loaded) and should have reached stability by now, or close to it. Thus the only edifices that might be responsible for the creation of a flexural bulge that would cause uplift in Santiago and São Nicolau are Fogo and Brava. However, we believe that the significant uplift experienced by Santiago and São Nicolau cannot be explained by far-field effects of multiple surface loading because: (1) both islands still lie within the flexural moat of the younger edifices of Fogo and Brava; (2) both islands are less than 100 km from nearby islands that do not exhibit contemporaneous uplift trends, but are, rather, stable [25]; (3) the amount of uplift is too large to be caused by multiple far-field surface loading. Furthermore, Brava exhibits evidence for $\sim$350 m of uplift [25] when this island should only have experienced subsidence or remained stable if only surface loading is considered.

In summary, only two other mechanisms may be responsible for the differential and significant uplift experienced by Santiago and São Nicolau (and Brava): (1) flexural rebound due to erosion and mass-wasting of the edifices and (2) flexural response to subsurface loading.

The flexural rebound associated with erosion and mass-wasting has previously been identified as a mechanism that contributes to island uplift [19–21]. However, the uplift component generated by this mechanism seems to be small: <110 m even considering a redistribution of 40% of the island volume and for a plate as thin as 25 km [62]. If we take a regional plate thickness (T_e) of $\sim$30 km [33, 34], over 100% of the island edifice would need to be eroded to explain the >450 m of uplift of Santiago assuming linearity of the effect. Furthermore, if any uplift occurred due to erosion and mass redistribution, it likely affected older islands rather than younger islands. Hence, we believe the contribution of erosion and mass-wasting rebound to the uplift of the Cape Verdes is very small.

Thus, only response to subsurface loading seems to provide a plausible cause for the reported uplift. The existence of subsurface loads—material added to the crust or its base by hotspot activity—has been invoked in many ocean islands to explain the particular geometry of their flexural moats and crustal structures [23] and surface uplift [63]. However, their role in modifying the shape of the flexed crust is still not clearly understood [23], and neither is the possibly different mechanism of underplating in fast-moving vs stationary plates. At first blush, one should expect that stability relative to the melting source that the long-term volcanism documents, coupled with a thick plate above the hotspot, would favor underplating, but some archipelagos in stationary environments do not show evidence of enhanced underplating [64]. Ali et al. [33] proposed the existence of subsurface loads beneath the Cape Verdes in order to explain the low-amplitude long-wavelength geometry of the moat and the outward tilting of the infill sequence. According to these authors, an uplift up to ~400 m is required to "unbend" the plate from the ~2,000 m inflection expected from island loading to the observed value of ~1,600 m. This value is remarkably similar to the total vertical displacement exhibited by Santiago: ~450 m. Furthermore, Ali et al. [33] proposed that the cause of their inferred uplift may be the existence of subsurface loads restricted to the islands' "footprint" and this seems to be supported by the crustal thickening inferred by Lodge and Helffrich [32]. The confinement of the subsurface loads to the islands' vicinities is also supported by the work of Pim et al. [34], since these authors did not find evidence for crustal thickening away from the islands. Thus, we believe the most likely cause for the reported uplift is related to differential magmatic additions below the edifices, but the details for this mechanism still need to be developed.

7.7 Conclusions

Our study of island uplift histories in the Cape Verdes reveals that:

1. Santiago has experienced a constant nearly-linear uplift trend in the last 4 Ma, at ~100 m/Ma (0.1 mm/a), with a resulting vertical displacement of >450 m.
2. São Nicolau seems to have experienced considerable uplift at 5.7–5.1 Ma, and at 0.5 Ma–present. The average uplift rate is smaller than Santiago's uplift rate: ~32 m/Ma (0.032 mm/a), with a vertical displacement of >260 m.
3. The data from our study and the general volcanic history inferred suggests that significant uplift movements may be synchronous with vigorous volcanic stages of island building.
4. The considerably distinct island uplift histories inferred from Cape Verde's geological record implies a local rather than regional mechanism as the cause of island uplift.
5. The reported uplift cannot be explained by far-field effects of multiple surface loading because: (1) both Santiago and São Nicolau islands lie within the flexural moat of younger edifices; (2) both islands are less than 100 km from nearby islands that do not exhibit contemporaneous uplift trends but are instead generally stable; (3) the amount of uplift is too large to be caused by multiple far-field surface loading. Likewise, the uplift cannot be explained by erosion and mass-wasting flexural rebound unless unrealistic volume wasting is considered.
6. The most likely cause for differential uplift might be related to varying amounts of magmatic additions beneath each island and restricted to the islands' vicinities.

References

1. Darwin, C. (1842). *On the structure and distribution of coral reefs*. London, UK: Ward Lock.
2. Scott, G., & Rotondo, G. (1983). A model to explain the differences between Pacific plate island-atoll types. *Coral Reefs,1*(3), 139–150.
3. Menard, H., & Atwater, T. (1969). Origin of fracture zone topography. *Nature,222*(5196), 1037–1040.
4. Parsons, B., & Sclater, J. (1977). An analysis of the variation of ocean floor bathymetry and heat flow with age. *Journal of Geophysical Research (Solid Earth),82*(5), 803–827.
5. Stein, C., & Stein, S. (1992). A model for the global variation in oceanic depth and heat flow with lithospheric age. *Nature,359*(6391), 123–129.
6. Menard, H., & Ladd, H. (1963). Oceanic islands, seamounts, guyots and atolls. *The Sea,3*, 365–385.
7. Brotchie, J. F., & Silvester, R. (1969). On crustal flexure. *Journal of Geophysical Research (Solid Earth),74*(22), 5240–5252.
8. Walcott, R. (1970). Flexure of the lithosphere at Hawaii. *Tectonophysics,9*(5), 435–446.
9. Menard, H. (1973). Depth anomalies and the bobbing motion of drifting islands. *Journal of Geophysical Research (Solid Earth),78*(B3), 5128–5137.
10. McNutt, M., & Menard, H. W. (1978). Lithospheric flexure and uplifted Atolls. *Journal of Geophysical Research (Solid Earth),83*(B3), 1206–1212.
11. Dietz, R., & Menard, H. (1953). Hawaiian swell, deep, and arch, and subsidence of the Hawaiian Islands. *The Journal of Geology,61*(2), 99–113.
12. Morgan, W. (1972). Deep mantle convection plumes and plate motions. *AAPG Bulletin,56*(2), 203–213.
13. Burke, K., & Wilson, J. (1972). Is the African Plate stationary? *Nature,239*(5372), 387–390.
14. McNutt, M. (1988). Thermal and mechanical properties of the Cape Verde Rise. *Journal of Geophysical Research (Solid Earth),93*(B4), 2784–2794.
15. Sleep, N. (1990). Hotspots and mantle plumes: Some phenomenology. *Journal of Geophysical Research (Solid Earth),95*(B5), 6715–6736.
16. Crough, S. (1978). Thermal origin of mid-plate hot-spot swells. *Geophysical Journal International,55*(2), 451–469.
17. Detrick, R., & Crough, T. (1978). Island subsidence, hot spots, and lithospheric thinning. *Journal of Geophysical Research (Solid Earth),83*(B3), 1236–1244.
18. Morgan, J., Morgan, W., & Price, E. (1995). Hotspot melting generates both hotspot volcanism and a hotspot swell. *Journal of Geophysical Research (Solid Earth),100*(B5), 8045–8062.
19. Rees, B., Detrick, R., & Coakley, B. (1993). Seismic stratigraphy of the Hawaiian flexural moat. *Bulletin of the Geological Society of America,105*(2), 189–205.
20. Filmer, P., McNutt, M., Webb, H., & Dixon, D. (1994). Volcanism and archipelagic aprons in the Marquesas and Hawaiian Islands. *Marine Geophysical Researches,16*(5), 385–406.
21. Schmincke, H.-U. (2004). *Volcanism* (1st ed.). Berlin: Springer.
22. Ribe, N. (2004). Through thick and thin. *Nature,427*(6974), 793–795.
23. Watts, A. (2001). *Isostasy and flexure of the lithosphere*. Cambridge, UK: Cambridge University Press.
24. Turcotte, D. L., & Schubert, G. (2002). *Geodynamics* (2nd ed.). Cambridge, UK: Cambridge University Press.
25. Ramalho, R., Helffrich, G., Schmidt, D., & Vance, D. (2000). Tracers of uplift and subsidence in the Cape Verde Archipelago. *Journal of the Geological Society 167*, 519–538.
26. Amante, C., & Eakins, B. W. (2009). ETOPO1 1 arc-minute global relief model: Procedures, data sources and analysis.*NOAA Technical Memorandum NESDIS NGDC-24,*19.
27. Torres, P., Silva, L., Serralheiro, A., Mendes, M., Macedo, J., & Gomes, A. (2002). Geologia da Ilha do Sal. Comunicações do Instituto de Investigação Científica Tropical.

28. Williams, C., Hill, I., Young, R., & White, R. (1990). Fracture zones across the Cape Verde Rise, NE Atlantic. *Journal of the Geological Society,147*(5), 851–857.
29. Monnereau, M., & Cazenave, A. (1990). Depth and geoid anomalies over oceanic hotspot swells: A global survey. *Journal of Geophysical Research (Solid Earth),95*(B10), 15–429.
30. Lancelot, Y., Seibold, E., Cepek, P., Dean, W., Eremeev, V., Gardner, J., et al. (1978). Site 368: Cape Verde Rise. In Y. Lancelot, E. Seibold, P. Cepek, W. Dean, V. Eremeev, J. Gardner, et al. (Eds.), *Initial reports of deep sea drilling project* (vol. 41, pp. 233–326). Washington: U.S. Government Printing Office.
31. Faugeres, J. C., Legigan, P., Maillet, N., & Latouche, C. (1989). Pelagic, turbiditic, and contouritic sequential deposits on the Cape Verde plateau (leg 108, site 659, Northwest Africa): sediment record during Neogene time. In *Proceedings of the Ocean Drilling Program, Scientific Results* (vol. 108, pp. 311–328).
32. Lodge, A., & Helffrich, G. (2006). Depleted swell root beneath the Cape Verde Islands. *Geology,34*(6), 449–452.
33. Ali, M. Y., Watts, A. B., & Hill, I. (2003). A seismic reflection profile study of lithospheric flexure in the vicinity of the Cape Verde Islands. *Journal of Geophysical Research (Solid Earth),108*(B5), 2239–2263.
34. Pim, J., Peirce, C., Watts, A., Grevemeyer, I., & Krabbenhoeft, A. (2008). Crustal structure and origin of the Cape Verde Rise. *Earth and Planetary Science Letters,272*(1–2), 422–428.
35. Holm, P., Grandvuinet, T., Friis, J., Wilson, J. R., Barker, A. K., & Plesner, S. (2008). An 40Ar-39Ar study of the Cape Verde hot spot: Temporal evolution in a semistationary plate environment. *Journal of Geophysical Research (Solid Earth),113*(B8), B08201.
36. Bernard-Griffiths, J., Cantagrel, J.-M., Alves, C., Mendes, F., Serralheiro, A., & Macedo, J. (1975). Geochronologie: Donnés radiometriques potassium-argon sur quelques formations magmatiques des îlles de l'archipel du Cap Vert CR. *Seances Academy of Sciences Series D,280*, 2429–2432.
37. Mitchell, J., Bas, M. L., Zielonka, J., & Furnes, H. (1983). On dating the magmatism of Maio, Cape Verde Islands. *Earth and Planetary Science Letters,64*(1), 61–76.
38. Plesner, S., Holm, P. M., & Wilson, J. R. (2002). 40Ar-39Ar geochronology of Santo Antão, Cape Verde Islands. *Journal of Volcanology and Geothermal Research,120*(1–2), 103–121.
39. Jørgensen, J., & Holm, P. (2002). Temporal variation and carbonatite contamination in primitive ocean island volcanics from São Vicente, Cape Verde Islands. *Chemical Geology,192*(3–4), 249–267.
40. Duprat, H., Friis, J., Holm, P., Grandvuinet, T., & Sørensen, R. (2007). The volcanic and geochemical development of São Nicolau, Cape Verde Islands: Constraints from field and 40Ar/39Ar evidence. *Journal of Volcanology and Geothermal Research,* 162(1–2), 1–19.
41. Foeken, J., Day, S., & Stuart, F. (2009). Cosmogenic 3He exposure dating of the Quaternary basalts from Fogo, Cape Verdes: Implications for rift zone and magmatic reorganisation. *Quaternary Geochronology,4*(1), 37–49.
42. Trindade, M. J., Mata, J., & Munhá, J. (2003). Petrogenesis of Quaternary magmatism in S. Vicente Island (Cape Verde). *Comunicações do Instituto Geológico e Mineiro,90*, 169–188.
43. Serralheiro, A. (1970). Geologia da Ilha de Maio (Cabo Verde), Junta de Investigações do Ultramar.
44. Paepe, P. D., Klerkx, J., Hertogen, J., & Plinke, P. (1974). Oceanic tholeiites on the Cape Verde Islands: petrochemical and geochemical evidence. *Earth and Planetary Science Letters,22*(4), 347–354.
45. Stillman, C., Furnes, H., Le Bas, M., Robertson, A., & Zielonka, J. (1982). The geological history of Maio, Cape Verde Islands. *Journal of the Geological Society,139*(3), 347–361.
46. Gerlach, D., Cliff, R., Davies, G., Norry, M., & Hodgson, N. (1988). Magma sources of the Cape Verde Archipelago: Isotopic and trace element constraints. *Geochimica et Cosmochimica Acta,52*, 2979–2992.
47. Davies, G., Norry, M., Gerlach, D., & Cliff, R. (1989). A combined chemical and Pb-Sr-Nd isotope study of the Azores and Cape Verde hot spots; the geodynamic implications. *Geological Society of London Special Publications,42*(1), 231–255.

48. Serralheiro, A. (1976). A Geologia da Ilha de Santiago (Cabo Verde). *Boletim do Museu e Laboratorio Mineralógico e Geológico da Faculdade de Ciências,14*, 157–369.
49. Silva, L., Serralheiro, A., Macedo, J., Gomes, A., & Torres, P. (1990). Carta Geológica de Cabo Verde, Ilha do Sal, na escala de 1/25000 (folhas 1-2), Edição do Instituto Investigação Científica Tropical/Instituto de Cooperação Económica.
50. Ubaldo, M., Silva, L., & Torres, P. (1991). Contribuição geológica e micropaleontologica para o conhecimento do "Complexo Eruptivo Antigo" da ilha do Sal, Arquipélago de Cabo Verde. *Garcia de Orta. Serviços Geológicos,14*(1–2), 9–14.
51. Torres, P., Silva, L., Serralheiro, A., Tassinari, C., & Munhá, J. (2002). Enquadramento geocronológico pelo método K/Ar das principais sequências vulcano-estratigráficas da Ilha do Sal - Cabo Verde. *Garcia de Orta Serviços Geológicos,18*(1–2), 9–13.
52. Macedo, J., Serralheiro, A., & Silva, L. (1988). Notícia Explicativa da Carta Geológica da Ilha de S Nicolau (Cabo Verde) na escala de 1:50000, Garcia de Orta. *Serviços Geológicos,11*(1–2), 1–32.
53. Davies, G. (1988). Ocean bathymetry and mantle convection, 1 Large-scale flow and hotspots. *Journal of Geophysical Research (Solid Earth),93*(B9), 10467–10480.
54. Martins, S., Mata, J., Munhá, J., Madeira, J., & Moreira, M. (2008). Evidências geológicas e geoquímicas para a existência de duas unidades estratigráficas distintas na Formação do Pico da Antónia (Ilha de Santiago, República de Cabo Verde). *Memórias e Notícias Universidade de Coimbra,3*, 123–128.
55. Bernoulli, D., Hottinger, L., Spezzaferri, S., & Stille, P. (2007). Miocene shallow-water limestones from São Nicolau (Cabo Verde): Caribbean-type benthic fauna and time constraints for volcanism. *Swiss Journal of Geosciences,100*(2), 215–225.
56. Miller, K., Kominz, M., Browning, J., Wright, J., Mountain, G., Katz, M., et al. (2005). The Phanerozoic record of global sea-level change. *Science,310*(5752), 1293–1298.
57. Kuiper, K., Deino, A., Hilgen, F., Krijgsman, W., Renne, P., & Wijbrans, J. (2008). Synchronizing rock clocks of Earth history. *Science,320*(5875), 500–504.
58. Hoffmann, D., Spotl, C., & Mangini, A. (2009). Micromill and in situ laser ablation sampling techniques for high spatial resolution MC-ICPMS U-Th dating of carbonates. *Chemical Geology,259*(3–4), 253–261.
59. Stirling, C., Esat, T., Lambeck, K., & McCulloch, M. (1998). Timing and duration of the Last Interglacial: evidence for a restricted interval of widespread coral reef growth. *Earth and Planetary Science Letters,160*(3–4), 745–762.
60. Esat, T., & Yokoyama, Y. (2006). Variability in the uranium isotopic composition of the oceans over glacial–interglacial timescales. *Geochimica et Cosmochimica Acta,70*(16), 4140–4150.
61. Grigg, R., & Jones, A. (1997). Uplift caused by lithospheric flexure in the Hawaiian Archipelago as revealed by elevated coral deposit. *Marine Geology,141*(1–4), 11–25.
62. Smith, J., & Wessel, P. (2000). Isostatic consequences of giant landslides on the Hawaiian Ridge. *Pure and Applied Geophysics,157*(6), 1097–1114.
63. Klügel, A., Hansteen, T., & Galipp, K. (2005). Magma storage and underplating beneath Cumbre Vieja volcano, La Palma (Canary Islands). *Earth and Planetary Science Letters,236*(1–2), 211–226.
64. Watts, A., Peirce, C., Collier, J., Dalwood, R., Canales, J., & Henstock, T. (1997). A seismic study of lithospheric flexure in the vicinity of Tenerife, Canary Islands. *Earth and Planetary Science Letters,146*(3–4), 431–447.

Chapter 8
Conclusions

The Cape Verde Islands are the result of long-term mid-plate volcanism and stand on a prominent bathymetric, gravity and geoid anomaly—all the classic features of a hotspot. The Islands lie on 120–140 Ma oceanic crust and are located on the western portion of the African plate, a plate that has barely moved in the last 30 Ma relatively to the hotspot reference frame [1]. For these reasons, the Cape Verde Archipelago is often regarded as the type-example of a hotspot-generated group of Islands in a stationary plate environment, making it an ideal place to study the effects of hotspot activity upon the Earth's surface and subsurface [2–4]. The stationary position of the plate relative to the hotspot reference frame implies that the archipelago is approximately stationary with respect to the melting source [3, 4]. This condition has a profound effect on the hotspot activity, as shall be discussed in this chapter. Furthermore, the stationary position also enhances the isostatic effects of hotspot activity, allowing a better separation between the individual mechanisms that contribute to island/seafloor uplift and subsidence.

One of the most prominent features associated with the Cape Verde hotspot is the Cape Verde Rise. This dome-shaped rise is 1200–1600 km in diameter and stands up to 1.9–2.3 km above the surrounding seafloor, making it the largest bathymetric anomaly in the oceans [2, 3, 5, 6]. The crustal and mantle structure of the swell has been inferred from receiver function analysis, and suggests an $\sim$80 km thick modified lithosphere with chemical buoyancy as the source of uplift [4]. This is in clear disagreement with the increased plate thinning expected from a thermal origin of the swell, or the underplating by warm material expected from dynamic support. Conversely, it suggests that the swell is supported by a cool buoyant root, probably constituted by depleted material left over from melting and melt extraction [4, 7]. The size and the axisymmetric geometry of the swell and swell root are greatly influenced by the plate's fixity relative to the melting source, because it allows the accumulation and spreading of material without the dragging induced by plate motion [3, 4, 7].

In order to better discriminate between different swell models, one needs to more accurately constrain the swell root geometry and investigate the evolution of the swell through time. Numerical forward modelling was used in this study to

R. A. S. Ramalho, *Building the Cape Verde Islands*, Springer Theses, DOI: 10.1007/978-3-642-19103-9_8, further test the hypothesis of a chemically buoyant root being the source of swell support in Cape Verde, and to better understand the possible swell evolution. Two different models were developed: a dynamic fluid-flow model coupled with purely isostatic uplift, and a static flexural model.

The fluid-flow model describes the shape and the evolution of a radially-symmetric spreading buoyant root and associated swell. Only numerical solutions yielding a swell root thickness between 60 and 80 km, a swell radius of 600–800 km, a swell height of 1.9–2.3 km, and a geoid anomaly between 8 and 12 m were accepted. Acceptable solutions correspond to the following parameters: (a) a density reduction between the fertile mantle and the depleted root ($\Delta\rho$) between 70 and 80 kg/m^3, which is within the range suggested by Pim et al. [8] for the centre of the Cape Verde swell; (b) a flux rate constant (α) between 0.70 and 1, suggesting conditions closer to a constant flux rather than a single release of material and also capturing the behaviour of an intermittent source; and (c) viscosities (η) close to 1.2×10^{20} Pa s. The model shows that the typical isostatic signature of a radially-symmetric spreading root is a rapid initial uplift that wanes with time until a stable vertical position is reached with almost no subsequent subsidence. The model also shows that, after being formed, the total swell height does not change significantly over time, but its radius increases at a slower rate. Consequently, seafloor located well away from the swell centre will be uplifted at a later stage than seafloor located near the swell centre. This pattern is compatible with the uplift inferred from the stratigraphy of several IODP sites located within the Cape Verde Rise. IODP sites 368 and 659, located near the swell centre, suggest a major uplift episode sometime around the Oligocene/Miocene transition or the Early Miocene [9, 10]. In contrast, site 12, which is located ~450 km away from the swell centre, suggests that uplift occurred later, around the Miocene/Pliocene transition [11, 12]. In a similar fashion, the evolution proposed by the model has clear implications regarding the vertical movements affecting the island edifices built on top and near the centre of the swell: only the older Islands, with an age that overlaps the swell's initial and main uplift episode, will record significant swell-related uplift; conversely, Islands built after the initial ~10 Ma of swell development will not record large scale swell-related uplift. This is also compatible with the uplift evidence found in the Islands' surface geology.

A static flexural model was also developed to better constrain the swell root shape. This model yields a swell root with 700 km in radius, 55 km in thickness, and a root density of 3245 kg/m^3 ($\Delta\rho = 85\,kg/m^3$), as the best fit solution for the observed bathymetry and considering a compensation depth ~80 km, $E = 100$ GPa and $T_e = 30$ km. The flexural model suggests, however, a swell root geometry different from the root described by the fluid-flow model, perhaps suggesting that the Huppert [13] model oversimplifies the episodic character of the source or possible density and viscosity heterogeneities.

The Cape Verde Islands do not form a single age-progressive linear chain. Instead, the Islands form two chains disposed in a rough semi-arc clustered slightly off-centre to the southwest of the Cape Verde bathymetric swell. There is no evident hotspot track but there is possibly an age progression in the southern chain, from east (older Islands) to west (younger Islands), as suggested by the age of the oldest exposed lavas. This age progression is also reflected in the geomorphology of the edifices: the older and razed volcanoes of the east contrast with the younger and prominent volcanoes of the west; the first are in a late post-erosional stage while the latter are in an early post-erosional stage, or still in the shield-building stage. The reasons for this age progression are still poorly understood, however. Likewise, the causes behind the particular geometry of the archipelago, with the Islands disposed in two different trends, are also still unclear.

The start of the volcanic activity associated with the Cape Verde hotspot is still poorly constrained because many of the Islands' basement complexes are not dated or are poorly dated. However, the available geochronology suggests that the volcanic activity probably started sometime in the Late Oligocene or Early Miocene. This seems to be the age of the oldest exposed lavas in the archipelago that are unequivocally associated with the hotspot activity [14], and coincides with the approximate timing of the major uplift episode of the Cape Verde swell. Not surprisingly, the oldest exposed lavas seem to correspond to the pillow-lavas and hyaloclastites of the Ancient Eruptive Complex of Sal ($\sim$26 Ma) which occur interbedded with Neogene pelagic carbonates [14–17]. These lavas denote significant uplift in the early stages of island building, and perhaps constitute the only exposed evidence for the initial uplift associated with swell development.

Judging by the volcanostratigraphy of the Islands and the available geochronology data, the volcanic lifetime of each individual island in Cape Verde is normally long, characterised by several periods of volcanism with decreasing volumetric output and intercalated with long (1–4 Ma) periods of quiescence. Likewise, more than one rejuvenated (post-erosional) volcanic stage may occur within the same island. These characteristics are probably another product of the archipelago's stationary position with respect to the melting source.

The Cape Verde's geological record spans 26 Ma and is remarkably rich in evidence for past sea-level positions. This evidence includes submarine volcanic units, lava deltas, marine sediments and wave-cut morphological features. Thus, through a detailed analysis of the volcanostratigraphy and morphology of each island, information can be extracted regarding the position of relative sea-level at various times of the Islands' evolutionary history. If geochronological data is available, relative sea-level may be compared with eustatic sea-level from a chosen eustatic curve and the vertical displacement between the two may be inferred and uplift/subsidence may be assessed. In this dissertation, a comprehensive analysis of relative sea-level change for each of the Cape Verde Islands was presented and discussed, based on field observations and stratigraphic information from the literature. The observations reported in this study show that differential vertical movements exhibit two end-member behaviours: (a) Islands with no significant uplift, like Santo Antão, São Vicente and Santa Luzia; (b) and Islands with significant uplift, like Santiago, São Nicolau and Brava. The latter Islands exhibit, respectively, positive vertical displacements in excess of 450, 260 and 350 m. Islands like Sal, Boavista and Maio seem to exhibit more complex histories with periods of both uplift and subsidence. Sal, in particular, shows evidence for an early uplift phase followed by episodes of lesser uplift and subsidence. In contrast with the Hawaiian Islands, however, none of the Islands seems to have endured significant post-shield subsidence. An assessment of the maximum vertical displacement experienced by each island is presented on Fig. 5.17 and shows well the disparities in uplift among the archipelago.

Among the Cape Verde Islands, Santiago and São Nicolau were particularly chosen for detailed study and sampling because: (a) both Islands occupy a central position in the archipelago; (b) both Islands exhibit a unique geological record, rich in palaeo-markers of sea-level; (c) the Islands have a similar age and evolutionary history, thus allowing the discrimination between different synchronous uplift histories. Uplift feature dating used a combination of U–Th (carbonates) and Ar–Ar (lavas) and allowed long-term vertical movement reconstructions. The results are shown in Fig. 7.6. Santiago seems to have experienced a constant nearly-linear uplift trend in the last 4 Ma, at ~100 m/Ma (0.1 mm/a), with a resulting vertical displacement of >450 m. In contrast, São Nicolau seems to have experienced considerable uplift at 5.4–5.1 Ma, and at 0.5 Ma—present. The average uplift rate is smaller than Santiago's uplift rate: ~32 m/Ma (0.032 mm/a), with a vertical displacement of >260 m. The reported uplift rates may be underestimated since the older palaeo-markers are considerably eroded, but nevertheless are comparable with other uplift rates found, for example, in the Hawaiian Islands: 0.032–0.1 mm/a in the Cape Verdes and 0.02–0.33 mm/a in Hawaii [18]. However, the latter study only encompassed the last 0.5 Ma whereas this study spans the last 5.7 Ma.

This study shows that both Santiago and São Nicolau exhibit evidence for a general uplift trend with large resultant vertical displacements that are seemingly synchronous with the volcanic activity that built the bulk of their exposed edifices. Moreover, both Islands seem to have been in fairly continuous uplift throughout their submarine and subaerial shield-building stages, extending well into their post-erosional stages. This shows that, contrary to what is known from the Hawaiian example, uplift may be synchronous with vigorous volcanic stages, despite the surface loading imposed by the extrusion of large volumes of magma during these stages. These observations also show that, unequivocally, the Islands exhibit different uplift histories. Considering that many of the nearby Islands do not show evidence for the same uplift trend, this confirms that the bulk of the uplift is differential among the Islands and is not a regional feature.

The causes for such differential uplift were investigated and several possible mechanisms were tested. According to this study, the uplift cannot be explained by far-field effects of multiple surface loading for several reasons. First, both Santiago and São Nicolau are too close to other Islands that do not synchronously exhibit a similar trend. If the uplift was fully flexural in origin, nearby Islands would rise according to the flexural model. Conversely, for the uplift to be fully flexural in nature, the contrast in uplift is too large for the small distance that separates these Islands. Second, the amount of uplift affecting Santiago and São Nicolau is too large to be explained by cumulative effects of multiple surface loading if one considers the geometry of the archipelago and the local flexural parameters ($T_e = 30$ km, $E = 100$ GPa). Third, both Islands still lie on the flexural moat of the younger edifices of Fogo and Brava, so subsidence should be expected, not uplift. Finally, the volcanostratigraphy of Brava suggests considerable uplift even though this island should only have experienced subsidence or stability if only surface loading is considered. Flexural rebound due to erosion and mass-wasting cannot explain the reported uplift unless unrealistic volumes of mass wasting took place. Thus, the most likely mechanism is probably related to varying amounts of intrusions at crustal level and restricted to the Islands's vicinities. A large portion of these intrusions must occur within the crust and neither above (within the volcanic edifice) or below: intrusions restricted to the base of the crust would create an upwards flexural response while intrusions above the crust would create a downwards flexural response, both situations incompatible in their pure form with the reported observations. The uplift may be partially flexural in nature, as suggested by the moat geometry [19], but its major component seems to be more local and not governed by a flexural mechanism.

The existence of subsurface loads has been invoked in many ocean Islands to explain the particular geometry of their flexural moats [20] and surface uplift [21]. According to Ali et al. [19] the Cape Verde flexural moat is shallower than expected and the moat infill tilts away from the Islands, suggesting that an uplift of $\sim$400 m occurred and modified the moat geometry. This value is similar to Santiago's $\sim$450 m uplift reported in this study. Furthermore, Ali et al. [19] suggested that the existence of subsurface loads restricted to the Islands' vicinities might be the origin of their reported uplift, a mechanism that is in general agreement with the crustal structure inferred by Lodge and Helffrich [4]. However, the differential uplift reported in this dissertation requires an uneven distribution of subsurface loads throughout the Cape Verde Islands, a condition also supported by the crustal structure inferred by Lodge and Helffrich [14]. The differential crustal thickening proposed by Lodge and Helffrich [14], up to 22 km, leaves plenty of scope for isostatic compensation associated with subsurface loads and suggests that intrusions at crustal level are common in the Cape Verdes. The stationary plate environment may contribute to this enhanced crustal thickening. Given these constraints, the uplift of the Cape Verde Islands might be largely explained by intrusions at crustal level. A direct correlation between the amount of uplift and the amount of crustal thickening is not evident, and is also not expected, because the exact position of these intrusions in the system has a profound effect on the nature and magnitude of the isostatic response (flexural or purely isostatic) and their occurrence may be erratic.

Evidence for Quaternary uplift is ubiquitous in the Cape Verde Islands. All the Islands, except Fogo which is too young and the western Islands from the northern chain that might have experienced some subsidence associated with the Santo Antão late loading episode (see Figs. 2.4 and 5.17), exhibit wave-cut surfaces, marine terraces and lava deltas at up to $\sim$100 m asl. These features suggest a recent uplift process that is regional in nature, probably acting at the swell's length scale. The reported Quaternary uplift is probably associated with a pulse of swell development, suggesting that this process is incremental and episodic in nature. Given that seafloor stratigraphy suggests a major uplift phase on the Cape Verde Rise before or around the Early Miocene [9, 10], an intermittent mechanism seems to be responsible for building the swell essentially instantaneously in an initial phase, with minor later incremental uplift. Of the three models of swell development, these features are consistent with thermal rejuvenation or spreading of melting residue, but not dynamic pressure. The dynamic pressure model implies a long-term pressure balance and is incompatible with incremental uplift because that would require incremental pressure increases. Since the thermal rejuvenation model is ruled out by the faster-than-average wavespeeds of the root, a depleted swell root model is favoured.

Thus, the long-term changes in island freeboard reported in this study suggest that at least two processes cooperatively act to raise the seafloor under ocean Islands (see Fig. 8.1). One process must act at the regional scale and a second

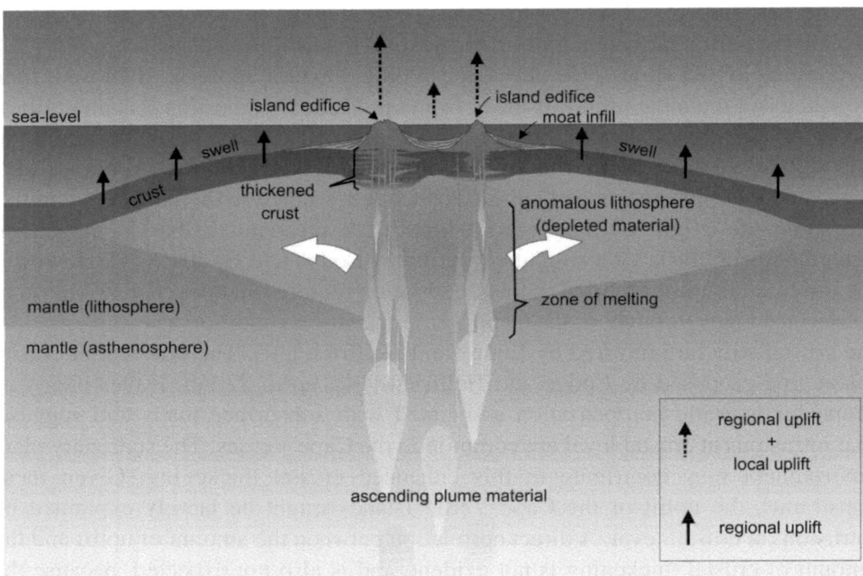

Fig. 8.1 Synoptic model for the causes of uplift in the Cape Verde region. Uplift is inferred to be caused by two main mechanisms: one acts at the regional scale and is associated with the episodic accumulation and spreading of a melting residue; another, more local and probably associated with magmatic additions to the crust, yields additional uplift. As decompressional melting occurs beneath the hotspot, efficient melt extraction leads to island volcanism and leaves behind a compositionally buoyant depleted melting residue that accumulates and spreads laterally, yielding regional uplift and creating the bathymetric anomaly. Swell uplift, like volcanism, seems to be episodic in nature and account for a portion of island uplift. Additional uplift seems to be yielded at a more local scale and is probably associated with significant crustal thickening by magmatic intrusions. The magnitude of locally-driven uplift is strongly influenced by the position of intrusions on the vertical column. Dimensions not to scale

process must locally act at the scale of an individual island to yield additional uplift. The first process is probably associated with swell development and favours the spreading of melt residue as the source of buoyancy for the Cape Verde hotspot swell; the second might be caused by cumulative crustal intrusions and may be synchronous with vigorous volcanic activity.

The Quaternary uplift is also responsible for maintaining the old and razed Islands of Sal, Boa Vista and Maio above sea-level. In the absence of uplift, and since these Islands barely had volcanism in their recent history, the edifices most probably would have been truncated by marine erosion by now, and would be in the guyot stage.

In a stationary plate environment there is no obvious mechanism for the waning of volcanism because the plate does not move away from the melting source. However, the enhanced crustal and lithospheric thickening suggested by the Cape Verde hotspot may provide the required mechanism for the decrease in volcanic activity; the stationary position, by enhancing thickening, also impairs the rise of magma to the surface and forces melting to occur at deeper levels.

8.1 Future Work

The results presented in this thesis suggest that the Cape Verde Islands have experienced significant differential uplift caused by a local rather than regional mechanism, probably associated with intrusions at the base of the edifice. However, many details remain poorly constrained and need further research. A brief overview on these details follows.

First, a better comprehension on the genesis of the Islands' basement complexes (Ancient Eruptive Complexes) is needed. This is of great importance because the basement complexes probably represent the initial stages of island evolution and bear witness to the vertical movements occurring during these stages. It is of particular relevance to determine whether these units comprise exclusively submarine products, or if they also include products of a subaerial nature. If only submarine products, these sequences may correspond to uplifted relics of the seamount stage. Conversely, if subaerial products occur underneath later submarine units, the basement complexes may correspond to initial cycles of island building that underwent subsidence contemporaneous with the edifice loading (evidence that is absent in the later volcanostratigraphic units). Likewise, it is of great importance to constrain the ages of these units. Information about their age is crucial to understand the possible migration of volcanism onset across the archipelago and, subsequently, the surface loading history associated with island building. However, the advanced weathering conditions and the profusion of dykes and sills that affect the majority of the Cape Verdes' basement complexes present real challenges to solid event reconstructions and geochronological studies.

The presence of Quaternary marine terraces and wave-cut surfaces above the relevant sea-level maxima in the majority of the Cape Verde Islands suggests a recent, regional uplift trend, possibly caused by a pulse in the Cape Verde swell. However, this uplift is still poorly constrained due to problems in the accurate dating of the relevant sea-level markers. To learn the details of this uplift is of key importance in the overall comprehension of hotspot swell evolution and island isostasy. Furthermore, a fine scale quantification of this uplift would provide a useful framework that is directly linkable with events that are better constrained, both in time and in space. Further research is thus needed to overcome the geochronological limiting factors and/or find alternative methods to date these markers. Cosmogenic exposure dating techniques (e.g. using ^{3}He and ^{53}Mn) seem particularly promising as potential ways to date wave-cut surfaces carved on submarine lavas or intrusive rocks (to avoid complex exposure histories).

In a similar fashion, further research is needed to understand the mechanisms behind the upheaval of the large pieces of sea-floor (pelagic sediments and MORB basalts) included in the basement complexes of Santiago and Maio. These mechanisms defy current ideas for island building because they imply a "bottom-up" (by doming and intrusions) rather than a "top-down" (by summit eruptions) edifice building process. The employment of geophysical methods might help in the reconstruction of the edifices' internal structure and assist in the comprehension of the processes of island building. Maio seems particularly suitable for a high resolution gravimetric survey because the contrast between the Mesozoic sediments and the Neogene lavas is great and because the island's low relief facilitates topographical corrections.

The loading history associated with the Cape Verde Archipelago is still crudely constrained, mostly due to our poor knowledge regarding the age of volcanic onset in each of the edifices. Nevertheless, the existence of a better, 3D flexural model would provide a valuable tool in the estimation of cumulative effects resulting from multiple surface loading.

Further research is also needed to develop a swell numerical model that can better incorporate the episodic character of hotspot activity and match the Cape Verde swell observational constraints. This model would have to incorporate the flexural parameters of the lithosphere, but at the same time be able to model the shape of the swell root in the temporal dimension. Additional information from seafloor stratigraphy, and regarding the timing of uplift for different positions along the Cape Verde Rise, would provide invaluable constraints on existing numerical models and swell processes.

References

1. Burke, K., & Wilson, J. (1972). Is the African plate stationary? *Nature, 239*(5372), 387–390.
2. McNutt, M. (1988). Thermal and mechanical properties of the Cape Verde Rise. *Journal of Geophysical Research (Solid Earth), 93*(B4), 2784–2794.
3. Morgan, J., Morgan, W., & Price, E. (1995). Hotspot melting generates both hotspot volcanism and a hotspot swell. *Journal of Geophysical Research (Solid Earth), 100*(B5), 8045–8062.

4. Lodge, A., & Helffrich, G. (2006). Depleted swell root beneath the Cape Verde Islands. *Geology, 34*(6), 449–452.
5. Crough, S. (1982). Geoid height anomalies over the Cape Verde Rise. *Marine Geophysical Researches, 5*(3), 263–271.
6. Monnereau, M., & Cazenave, A. (1990). Depth and geoid anomalies over oceanic hotspot swells: A global survey. *Journal of Geophysical Research (Solid Earth), 95*(B10), 15429.
7. Lodge, A. (2006). *Seismic constraints on swell formation beneath the Cape Verde Islands*. Ph.D. thesis, University of Bristol.
8. Pim, J., Peirce, C., Watts, A., Grevemeyer, I., & Krabbenhoeft, A. (2008). Crustal structure and origin of the Cape Verde Rise. *Earth and Planetary Science Letters, 272*(1–2), 422–428.
9. Lancelot, Y., Seibold, E., Cepek, P., Dean, W., Eremeev, V., Gardner, J., Jansa, L., et al. (1978). Site 368 Cape Verde Rise. In Y. Lancelot, E. Seibold, P. Cepek, W. Dean, V. Eremeev, J. Gardner, L. Jansa, D. Johnson, V. Krasheninnikov, U. Pflaumann, J. G. Rankin, P. Trabant, D. Bukry (Eds.), *Initial reports of deep sea drilling project* (Vol. 41, pp. 233–326). Washington: US Government Printing Office.
10. Faugeres, J. C., Legigan, P., Maillet, N., & Latouche, C. (1989). *Pelagic, turbiditic, and contouritic sequential deposits on the Cape Verde plateau (leg 108, site 659, Northwest Africa): Sediment record during Neogene time.* In Proceedings of the Ocean Drilling Program, Scientific Results, Vol. 108, pp. 311–328.
11. Peterson, M., Edgar, N., von der Borch, C., Cita, M., Gartner, S., Goll, R., et al. (1970). Site 12. In: M. Peterson, N. Edgar, M. Cita, S. Gartner, R. Goll, C. Nigrini, C. von der Borch (Eds.), *Initial reports of deep sea drilling project* (Vol. 2, pp. 249–303). Washington: US Government Printing Office.
12. Peterson, M., Edgar, N., von der Borch, C., & Rex, R. (1970). Cruise leg summary and discussion. In: M. Peterson, N. Edgar, M. Cita, S. Gartner, R. Goll, C. Nigrini, C. von der Borch (Eds.), *Initial reports of deep sea drilling project* (Vol. 2, pp. 413–427). Washington: US Government Printing Office.
13. Huppert H. (1982). The propagation of two-dimensional and axisymmetric viscous gravity currents over a rigid horizontal surface. *Journal of Fluid Mechanics, 121*(1), 43–58.
14. Torres P., Silva L., Serralheiro A., Tassinari C., & Munhá J. (2002). Enquadramento geocronológico pelo método K/Ar das principais sequências vulcano-estratigráficas da Ilha do Sal—Cabo Verde. *Garcia de Orta, Serviços Geológicos, 18*(1–2), 9–13.
15. Silva, L., Serralheiro, A., Macedo, J., Gomes, A., & Torres, P. (1990). Carta Geológica de Cabo Verde, Ilha do Sal, na escala de 1/25000 (folhas 1-2), Edição do Instituto Investigação Científica Tropical/Instituto de Cooperação Económica.
16. Torres, P., Silva, L., Serralheiro, A., Mendes, M., Macedo, J., & Gomes, A. (2002). Geologia da Ilha do Sal. Comunicações do Instituto de Investigação Científica Tropical.
17. Ubaldo, M., Silva, L., & Torres, P. (1991). Contribuição geológica e micropaleontologica para o conhecimento do "Complexo Eruptivo Antigo" da ilha do Sal, Arquipélago de Cabo Verde. *Garcia de Orta, Serviços Geológicos, 14*(1–2), 9–14.
18. Grigg, R., & Jones, A. (1997). Uplift caused by lithospheric flexure in the Hawaiian Archipelago as revealed by elevated coral deposit. *Marine geology, 141*(1–4), 11–25.
19. Ali, M. Y., Watts, A. B., & Hill, I. (2003). A seismic reflection profile study of lithospheric flexure in the vicinity of the Cape Verde Islands. *Journal of Geophysical Research (Solid Earth), 108*(B5), 2239–2263
20. Watts, A. (2001). *Isostasy and flexure of the lithosphere*. Cambridge: Cambridge University Press
21. Klügel, A., Hansteen, T., & Galipp, K. (2005). Magma storage and underplating beneath Cumbre Vieja volcano, La Palma (Canary Islands). *Earth and Planetary Science Letters, 236*(1–2), 211–226.

MIX
Papier aus verantwortungsvollen Quellen
Paper from responsible sources
FSC® C105338

If you have any concerns about our products,
you can contact us on
ProductSafety@springernature.com

In case Publisher is established outside the EU,
the EU authorized representative is:
Springer Nature Customer Service Center GmbH
Europaplatz 3, 69115 Heidelberg, Germany

Printed by Libri Plureos GmbH
in Hamburg, Germany